LUCIFER'S LEGACY

LUCIFER'S LEGACY

The Meaning of Asymmetry

FRANK CLOSE

DOVER PUBLICATIONS, INC.
Mineola, New York

Bibliographical Note

This Dover edition, first published in 2013, is a revised republication of
the work originally published in 2000 by Oxford University Press, Oxford.

Library of Congress Cataloging-in-Publication Data

Close, F. E.
  Lucifer's legacy : the meaning of asymmetry / Frank Close
    pages cm
  Originally published: Oxford ; New York : Oxford University Press, 2000.
  Includes index.
  ISBN 978-0-486-49167-7 — ISBN 0-486-49167-6 1. Symmetry (Physics)
I. Title.
  QC174.17.S9C56 2013
  530—dc23

                                                           2013014078

Manufactured in the United States by Courier Corporation
49167601   2013
www.doverpublications.com

For Katie and Lizzie: the Real Legacy

Acknowledgments

At the end of a film, the credits roll for several minutes. Some-one must have kept a checklist of every person that chanced upon the set so that they can be listed alongside the codenames of theatre, such as BestBoy, Grip, and other characters that seem to have spilled over from Lewis Carroll. I regret that I have not been so efficient. Anecdotes have come from people who have asked some perceptive question or made a comment after talks that I have given and of whom I have no record. In addition, many dis-cussions with speakers at the British Association for the Advance-ment of Science, or with colleagues at the European Centre for Particle Physics CERN in Geneva, and in the Rutherford Appleton Laboratory are too numerous to mention other than collectively.

However, among these there are a few that I would like to acknowledge specifically. Talks on symmetry by Denys Wilkin-son, Ian Stewart, and Chris Llewellyn Smith stimulated my inter-est in this area; I enjoyed meeting with Henning Genz and sharing ideas on how to popularise the Higgs mechanism and I am also indebted to Peter Higgs for copies of his original manuscripts and anecdotes. Peter Kalmus' insights in his talk on the Forces of Nature also deserve mention as do the help I received from Lewis Wolpert, Chris McManus, Steve Laval, Dahlia Zaidel, Charles Stirling, Peter Watson, and Ken Zetie on the bizarre role of mir-ror asymmetries in atoms, chemistry, and living organisms. Neil Calder, Mick Draper, and Rolf Landua at CERN and Bryson Gore at the Royal Institution have helped in providing some of the conceptual illustrations and Tim Radford and David Newnham at *The Guardian* have helped my thoughts on antimatter to mature. Good editors are essential and I am indebted to Carole Sunderland, and also to Susan Harrison and Beth Knight, at Oxford University Press, for their role in bringing Lucifer's Legacy to life.

Contents

Chapter 1

Lucifer

"Headless body found in topless bar"
(USA newspaper headline)

The world is an asymmetrical place full of asymmetrical beings. If the Creation had been perfect, and its symmetry had remained unblemished, nothing that we now know would ever have been. There would have been no you to read this book, nor me to have written it; there would have been no Paris in the spring, and no Tuileries Gardens. So I would not have come across the headless body—the chance event that started me wondering on the accident, or design, that has created life out of arid equations.

Lest you make this for a seamy pot-boiler, or even Hercule Poirot–style murder mystery, I should make clear at the outset that the body in question was made of stone, its head lying in the gravel at the base of a plinth which bore the legend "Lucifer." Had it been other than in the Tuileries I would probably have passed by without giving a second glance, but the gardens are beautiful, laid out with mathematical precision. I paused and looked again at headless Lucifer. Its disfigured presence in the midst of an otherwise all pervading perfection was as profound as the unresolved chord that ends Bach's St Matthew Passion and seemed like a metaphor for existence.

The spring day had started slowly as I had come to Paris by train from London. The carriages had meandered through southern England, as if to give the passengers time to appreciate the picture postcard views of Kent, before speeding through the featureless landscapes of northern France, impatient for the graceful architecture of Paris, among which is one of the most remarkable perspectives in Europe. From the midpoint of the Arc de Triomphe the line of sight along the Champs Elysées leads first to the obelisk at the centre of the Place de la Concorde and then runs the length of the Tuileries Gardens to the open arms of the Louvre. Where this line passes through the Tuileries it has been used like a mirror such that with draughtsmanlike precision the two halves of the gardens are perfect reflections of one another (Fig. 1.1).

To experience the symmetry to the full, first stand at the western end in the Place de la Concorde looking in the direction of the Louvre at the far end of the Tuileries. Within the park and a hundred metres to your right is the Orangerie, a hothouse erected by Napoleon III and now used for art shows; its mirror image, the same distance on your left, is the Musée du Jeu de Paume, Napoleon's tennis court. Identical mirrored paths connect the two buildings to your position at the park's entrance. On your immediate left you will see a high stone wall, curved with its concave side towards you; to the right of you is another wall, its arc a perfect mirror of its partner. The pair are a miniature imitation of the curved collonades that caress your arrival before the entrance of St Peter's in Rome. Whereas statues of angels decorate Michaelangelo's creation, the entrance to the Tuileries holds sinister statues of philosophers, gods, and dead Frenchmen in its embrace; to your left and to your right the guardians stand in perfect symmetry. The effect though is similar. The pair of symmetric concave curves are like a mother's arms taking her child; they welcome and encourage you to enter within.

The central avenue leads proudly from the far side of the curve, diametrically opposite you. This gravel carriageway defines the axis of the imaginary mirror, identical trees and flower-beds bordering it to left and right. Were you to rest briefly on one of the benches alongside the path and survey the beauty of the gardens, you would find the view obscured by a corresponding bench across the way. To every statue on one side there is a statue on the other, to every tree a tree, to every flower garden on the north

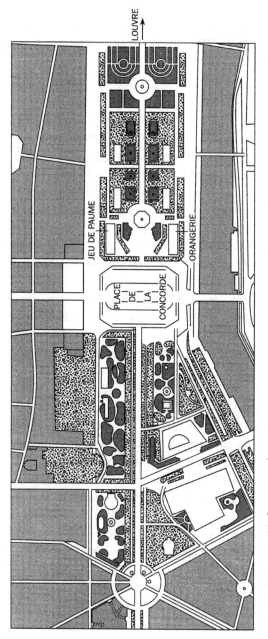

Fig. 1.1 Map of the Tuileries Gardens

side there is another planted at the same distance to the south. A water fountain sprays from the mouth of a nymph who is gazing soulfully at its clone, forever separated by twice the distance to the centre line of the park.

And so it went on until I saw the headless devil twenty metres down a side path. I knew that behind me, as yet unseen, would be a mirror image of this path that would lead to a correspondingly positioned plinth and fiendish statue. I half expected that this too would be broken, so preserving the symmetry of the park, but when I turned and looked I saw that its diabolic twin grinned from its plinth as it had done since its creation. In the entire gardens the designer symmetry was perfect with the sole exception of headless Lucifer.

The symmetry of the Tuileries Gardens and its interruption by the disfigured devil are metaphors for our grander perceptions of the natural world. Symmetry is fascinating and appealing; scientists seek it in their data and incorporate it in their theories, ironically even when there is no immediate evidence for it. Perhaps the most arcane example of this concerns the nature of matter and the fabric of existence embodied in the current cosmophysical description of Creation.

There is direct evidence that the stuff of which we are made is but half of a symmetric whole. Scientists speak casually of antimatter, the faithful opposite matter, the symmetry so perfect that when any particle of matter meets its mirror antiparticle, mutual annihilation occurs. It is the romance of this mutual suicide pact and the accompanying burst of energy that has made antimatter so beloved by science fiction writers and the chosen power source of the Starship *Enterprise* in "Star Trek." Physicists at CERN, the European Centre for Particle Physics in Geneva, can even watch this happen, confirming time and again the vulnerability of matter for antimatter. They also see the converse, where a large enough concentration of energy can coagulate into the two forms of substance: matter, as we know it, and its mirror image, antimatter.

So precisely balanced are these twins, so symmetrical their behaviour, that it has become the dogma of the new theology (or at least the one that is popular among cosmologists) that the two were made equally in the Creation. In the beginning, they say, there was no time, no space, no substance of any kind; this is a modern version of "there was darkness on the face of the void." Then came a burst of energy: "let there be light and there was light." Where did it come from? You may well ask. I don't know and nor, in my opinion, does any scientist with certainty. Questions concerning existence "before" this singular happening are wracked with philosophical debate as to whether they are even meaningful: what means "before" if there was no space nor time? Some popular descriptions seem content to portray a will-o'-the-wisp universe which erupted as a quantum fluctuation out of nothing. Maybe it did, but if so then I would feel compelled to ask why it bothered. Questions relating to the "spontaneous" appearance of that first flash of searing heat that we loosely call the Big Bang begin with "why" and, as such, are beyond (current) experimental scientific enquiry.

Much of the sometimes confused debate about science "versus" religion fails to distinguish between the "why?" and "how?" varieties of question. **Why** the void erupted into light is for others to debate; **how** our present material universe emerged from that light of creation is within science's province. Answers come from experiment—a crucial feature that distinguishes our modern scientific saga from other myths. Thus the CERN experiments on the production of the simplest particles and antiparticles out of radiant energy have inspired the current theories on the emergence of matter and antimatter during the Big Bang. A perfect Creation, with its symmetry untainted, would have led to matter and antimatter in precise balance and a mutual annihilation when in the very next instant they recombined: a precisely symmetrical universe would have vanished as soon as it had appeared. Such a uniform cosmic soup could hardly have led to the asymmetrical universe that we are a part of today where antimatter appears to be all but absent.

The current theory is that Creation was barely completed before something interceded; the perfection where the essence of every atom of substance had been counterbalanced by a precise anti-partner was lost forever. This act degraded the symmetry between matter and antimatter, with the result that after the great annihilation, a small portion of the matter was left over. Those remnants are what have formed us and everything around us as far as we can see. We are the material rump of what must have been an even grander Creation.

The flawed opus is what we are left with and, but for which, we would not be. Thus did the broken statue of Lucifer, spoiling the balance of the Tuileries Gardens, so brusquely remind me that the real world is full of asymmetrical features. The sly smirk on the Devil's face seemed to be one of victory as if humans and even the existence of the entire material universe are permanent legacies of blemishes introduced somehow during the Creation.

This disruption to the great design set me wondering about the multitudes of natural asymmetries that seem to have been necessary for human life to have emerged. Matter defeating antimatter was a necessary step for there to be anything at all, but this alone was not enough. Had that been the end of it, the material universe would have been merely a bland plasma of particles with no periodic table of the elements needed for life nor a solid earth to be the factory for its construction.

The simplest element, hydrogen, formed first, and the force of gravity collected it into the vast clumps that are stars, such as our sun. Had gravity been the only force at work, that would have been the end of the story: elemental pieces of hydrogen, falling in on one another, swirling into the vortices of black holes, and extinction. A simple implosive story perhaps, but with no sentient beings to record it. However, nature differentiated other forces that can transmute the elements, producing the profound version of a material universe that we are privileged to have evolved in.

With hydrogen as the fuel, the stellar cooker first produces helium and then mixes the heavier elements such as carbon, nitrogen, and oxygen—so necessary for life. The sun has been in the first stage, burning hydrogen, for five thousand million years and radiating sunlight across a hundred million miles of space to our planet through that time. It is this warmth that has energized the chemical and biological processes of life, which have in turn needed that vast timespan to evolve complex human systems from primeval DNA.

Here, once more, asymmetry has been necessary. The warmth from the sun, the radiant glory of the electromagnetic force, has vibrated all the way from that distant ball of fire, whereas the force involved in the transmutation of hydrogen in the sun has its sphere of influence smaller than the dimensions of an individual atomic particle. Its strength is much weaker than that of the electromagnetic force. This enfeeblement is what has enabled the sun to survive; had it not been like this, had the force driving the solar furnace been as powerful as the electromagnetic force, all of the solar fuel would have been exhausted within five hundred thousand years—far too brief a time for life on earth, or anywhere, to have emerged. This separation of the electromagnetic force and its aptly named "weak" sibling is but one of the critical asymmetries that has been necessary for our existence.

The structure of the atomic elements also is lopsided. Biology, chemistry, and life are the result of electric currents—coursing through the nervous system, changing food into energy, building our bodies and the very fabric of the planet. It is the journeying of the little electrons, the carriers of electrical charge, that determine everything that we experience. The individual atoms consist of these negatively charged electrons swarming around a static, bulky, positively charged nucleus. All but one of the two thousand parts of the mass of an atom reside in this central nucleus, while the tiny electrons flow from one atom to another; liberated as current they flow through wires and power modern industry; agitated by electric fields they radiate electromagnetic waves. It is these negative charges that communicate and drive the biochemical processes in living things while the positives, too heavy to be easily stirred, tend to stay at home and form the templates of solidity. This asymmetry in mass is crucial for the structure of materials.

However, this alone appears to be insufficient for life. Life appears to thrive on mirror asymmetry, a distinction between left and right in the basic structures of organic molecules. Let me expand on this now, as it will be central to our story.

The positive seeds with their negative captives form atoms and molecules. There are simple ones such as water; more complex examples such as amino acids, proteins, and DNA; and others created by human ingenuity, such as plastics, ceramics, and drugs. Most of these have shapes that differ from their mirror images. Superficially identical in all respects but for the interchange of left and right, one might have reasonably expected that both forms would be equally abundant in nature. However, it is not so; life is mirror asymmetric. This is not simply a matter of there being more right handers than left, or even of our heart and stomach being found, usually, on our left side. The amino acids and molecules of life in one form have the ability to know that they exist and to be cogniscent of the universe; their mirror images are inorganic, lifeless. Life chooses one form while the mirror image is rejected. The body may happily digest a substance in one of the two mirror forms as food while excreting its mirror image unused, or worse, be poisoned by it. How and why has this mirror asymmetry emerged?

The deeper one looks, the more asymmetry becomes apparent and seemingly necessary for anything "useful" to have emerged. Without asymmetry and structure, the universe would have been bland. Have we convinced ourselves that the Creation was perfect on nothing more than wish fulfillment, as evidence of imperfection and asymmetry is all around us and even within us? I wonder whether the multitudes of asymmetries are the proof that we are the end products of chance and that philosophers and scientists have created a quasi-religious parable of symmetry that is obscuring the real explanation. Or was there indeed a perfect symmetrical scheme which included some wonderful single ingredient, yet to be identified, from which all the asymmetries for life have spontaneously emerged?

The focus of much current research is to understand how nature hides symmetry, producing structured patterns out of underlying uniformity. The quest to find the answer and possibly the singular source of all asymmetry, the reason for form and existence, began in the Tuileries Gardens and culminated in this book.

The story divides effectively into three parts. First, the mystery is introduced in Chapters 2 to 4. Then, in Chapters 5 to 7, we meet the forensic tools that have been central in solving it. These chapters tell of the discoveries, a hundred years ago, of X-rays, radioactivity, and of the structure of the atom from which have emerged the modern profound insights into the origins and evolution of life and the universe. These three chapters are self-contained, primarily of historic interest and provide a background for the main story. Chapters 8 to 13 put these forensic tools to work to reveal what scientists currently believe to be the source of structure and asymmetry in nature, and describe how they are now solving the puzzle during the first years of this new century.

Chapter 2

Symmetry at large

Travel from Europe to Australia or from North to South America and you will (nearly) have turned yourself upside down. As children, when we first discover this, we tend to wonder if Australians have difficulty staying on the ground or if, permanently head over heels, eyes bulge with the blood pressure. Later, we may meet our one-time contemporaries from "down under" and discover that they grew up with the same concerns about us. By then we know that gravity pulls everything towards the centre of the earth and that citizens from the antipodes believe themselves to be as upright as those in the north.

Nonetheless, we **are** inverted relative to one another and, as I discovered on my first visit to the Southern hemisphere, it is possible to tell—at night. It was January and crystal clear in England. As the perfect golden circle of the sun set in the west, Venus and Jupiter began to appear, like mirrors reflecting the sunlight, followed by the full moon rising in the east and Orion dominating the sky to the south. Orion, the hunter, is one of the easiest constellations to recognize, with Betelgeuse and Rigel at his shoulders and feet, and a trio of stars forming his belt from which further stars appear to hang like a dagger. Twenty-four hours later I was in southern Africa. What appeared to be an identical sun to that of the previous evening duly sank below the horizon, but the night sky that replaced it was quite different. Of course it contained many stars such as the Southern Cross that were unknown to me, obscured by the earth to northern eyes, but the constellations over the equator, such as Orion, were present and yet somehow unfamiliar. Orion's dagger had turned into the phallus of a satyr; the face in the moon looked seriously ill, with its eyes below its mouth. Then I realized what had happened: I was viewing them inverted, as if standing on my head which, in a sense, I was.

While the irregular patterns of the constellations or the defining features in the moon help to show which way up we are by night, it is not so immediately obvious by day. The sun does cross the sky from right to left viewed from southern Africa, rather than from left to right in the Northern hemisphere. However, that takes a while to discern (unless you are an aficionado of sundials and notice that those in the Southern hemisphere are mirror images of their European counterparts) and at any moment the image of the southern sun looks pretty much like its northern form—a feature-less circle. Viewed in isolation of the horizon, there is nothing to define which way is "up" on the sun. It is symmetrical, presenting the same image to all orientations.

Of course, I had no doubt that I was in South Africa rather than England as the winter snow had been replaced by the swimming pools of summer. The earth, tilting up and down through the year, is in January pointing the southern half up at the sun while the north gets mere glances. Greenland is always icy, though Iceland is green, at least in June when the sun is above the horizon for 24 hours a day. In January, by contrast, it is the turn of the Antarctic to enjoy continuous daylight while the northern reaches suffer the long night of the Arctic winter. At any one spot, from pictures taken at midday during the year, we could tell the season by the changing height of the sun in the sky. However, the sun itself would appear to be the same in each image. This is quite remarkable as we get a different perspective on the sun each day. During the course of a year the earth carries us on a grand tour around the sun. Between January and July we travel halfway round and view the sun from opposite faces, while in April and October we are viewing it effectively from the sides. Every day the sun's shape appears as a constant perfect circle even though we are viewing it from all directions. This is because the sun is a sphere, presenting circles of the same size to us from whichever direction we look.

The sphere permeates the cosmos. Not only is our sun spherical but other stars are too, as are the planets and the moon (apart from irregular surface features). Out of the infinite variety of shapes and forms that might have been, all of these heavenly objects have chosen the sphere. It is not only the individual stars that are like this: the entire cosmos appears to have the overall symmetry of a sphere, where no single direction in space is favoured over any other. Our sun is but one of billions of stars in our galaxy, the Milky Way, which belongs to a cluster of galaxies. Beyond this cluster, deep in space, are other clusters of galaxies. The entire cosmos is the result of a Big Bang which marked the start of space and time some fifteen thousand million years ago. These clusters of galaxies are rushing away from one another in all directions as a result of that long-ago explosion. Imagine that you cut an enormous slice through the universe. It does not matter which direction you make the slice, the gross features of the universe are the same on one side as on the other: the density of galactic clusters is the same, their outward rushings are the same, and the varieties of stars within them are the same. The universe as a whole has the symmetry of a sphere.

The basic fabric of space, at least as perceived by human eyes, cares naught for any one direction more than another. This is a dramatic observation. The fact that the entire cosmos has a common feature implies that there is something deeply encoded in the laws of nature that makes it like this.

A sphere is the natural shape that forms when the only force attracting the constituents to one another cares only about the distances between them and not the direction. Gravity is an example: the attraction is in proportion to the masses of the constituents, be they hydrogen atoms or whole stars, and dies off fourfold for every doubling of the distance (the "inverse square law" of Isaac

Newton), but the direction is irrelevant. Individual pieces of hydrogen attract one another from all directions equally and the resulting cloud is a sphere. As they bump into one another the agitation heats them until they begin to glow. A star is born. Our sun, and all stars, started life as a spherical ball of hydrogen.

This illustrates why appreciation of symmetry can be so profound in natural philosophy. The realization that nature treats all directions uniformly, symmetrically, immediately leads us to expect that spherical shapes will be the natural order. The gross structure of the cosmos and the individual stars bear witness to that. However, not everything is spherical, though other patterns of symmetry arise. Humans for example are certainly not spherical, though we are roughly mirror symmetric from left to right, if not from top to bottom. An important part of the reason is that stars are surprisingly simple forms of matter whose shape is determined essentially by gravity alone; the structure of the human body, by contrast, is a result of complex electric and magnetic interactions among the atoms and molecules within. It is harder therefore to understand the structure and dynamics of living creatures than of

Fig. 2.1 Isaac Newton on an old British pound note. The planetary orbits are shown as ellipses whose shapes are greatly exaggerated: in the actual solar system the orbits are much nearer to circles. In reality, the sun is at the focus of the ellipse, not at the centre as shown on this banknote.

stars. Nonetheless, symmetry and asymmetry provide important clues, as we shall later see.

What is symmetry?

We have talked about symmetry without really saying what it is. We tend to use it to describe things that are well balanced, exhibiting regular forms or patterns—a collection of similar features that blend harmoniously providing a sense of beauty in the whole. However, "beauty is in the eye of the beholder," so we need to define symmetry more precisely than this. The bilateral symmetry of right and left which is so obvious in the human body, and especially in the face, is a good example. While one may debate whether this face or that one is more or less beautiful, it is possible to define precisely whether one or the other is more or less symmetric. If an object has bilateral symmetry, then it will appear identical to its mirror image; if an object differs from its mirror image then it does not have bilateral symmetry.

This is a particular example of the general concept of symmetry as used by mathematicians. If you view an object from a different perspective, such as rotating it, turning it over, or looking at its mirror image, then it is said to be symmetric if what you see appears to be the same as the original. Some examples will make this clear and also illustrate different classes of symmetry.

Imagine the surface of a huge lake, stretching as far as the eye can see; there is not a breath of wind and the surface is perfectly smooth. Move a hundred metres in any direction and the lake looks exactly the same. This is known as symmetry under translation. There is nothing to show which direction is which (at least as far as the surface of the lake is concerned; the position of the sun would give the game away in general of course): rotate through 30 degrees and the view is identical. We say that the surface is symmetric under rotation. Now drop a stone into the lake and ripples spread out in perfect circles from the impact point. The point where the stone enters breaks this translation symmetry but rotational symmetry remains: turn around, face any direction, and the ripples look the same.

So much for one stone. Now suppose that two stones were dropped simultaneously at different places, for example, one due east of the other. This breaks the rotational symmetry, since if we face east we are looking directly along the line of the two splashes whereas if we rotate through 90 degrees we are looking orthogonal to it. No longer are the views in all directions identical: rotational symmetry has been broken. However, there is still one special angle of rotation where symmetry survives. Turn completely around through 180 degrees so that instead of facing east you are now facing west. The view does remain the same in this particular case.

The illustration becomes more interesting when more stones are involved. Suppose that we had dropped three stones at points forming the corners of an isosceles triangle: rotation through 120 degrees or 240 degrees would preserve the view. As a more extreme example suppose that we dropped 12 identical stones at the 12 points of the clock. Ripples spread out from each forming a

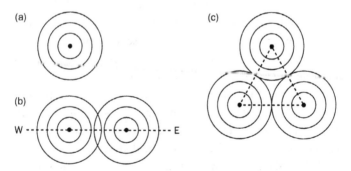

Fig. 2.2 Ripples from stones and symmetry: (a) rotate through any angle and the image remains the same; (b) the image is unchanged when rotated through $180° = 360°/2$ (c) the image is preserved when rotated through $120° = 360°/3$ or $240°$.

complicated but regular pattern as they cross one another. Rotate through 30 degrees, 60 degrees, or multiples thereof, and the pattern of ripples will appear identical. There is no rotational symmetry but there is a discrete symmetry for rotation through these special angles. If you had been shown a picture of the merging ripples you would intuitively have recognized its pattern, its symmetry. The rotations that leave the image unchanged are how the mathematicians define the underlying symmetry.

Once you appreciate that our intuitive feeling for symmetry has a strict underlying definition, then you will be able immediately to see the catch in the old canard "why do mirrors invert left and right but not top and bottom?" This question is actually a fraud, inviting an answer under the implicit, and illegal, assumption that the question refers to a real event. Who says that mirrors DO invert left and right? Suppose that humans were not so superficially bilaterally symmetric, for example that we were harlequins whose right half was red and our left lilac. Furthermore, suppose that everyone was like that; that no one existed with reddened left and lilac to the right. In such a world when you looked in a mirror you would see an image of a creature that does not exist in the real world. There would no longer be bilateral symmetry which shows that it is the nature of the shape, not the mirror, that determines the symmetry. "So what is it that a mirror really does?" I imagine you asking in frustration. We shall deal with this in Chapter 3.

Asymmetries

An essential part of the complexity of life is that many forces and interwoven components are competing and cooperating in building up the creature, in contrast to the case of a star: biology is intrinsically more complex than astrophysics. However, even when there is only gravity at work, the simplest disturbance can hide the "natural" spherical symmetry, transforming it into other patterns with their own symmetrical forms.

In a telescope it is possible to see entire galaxies of stars, where billions of suns are gathered together through their mutual gravitational attraction. Some galaxies are themselves spherical overall with more stars near the centre than the periphery. When I viewed one through a telescope, across the immensity of the intervening space I felt that I could almost "see" gravity pulling the individual stars towards the centre from all directions equally. Here on the galactic scale is a beautiful example of the spherical symmetry at work.

However, most galaxies are not spherical. For example, our own, the Milky Way, is relatively flat, like a disc. Essentially all of the stars that we see with the naked eye belong to our galaxy. Looking up or down out of the disc we see relatively few stars between us and deep intergalactic space, whereas if we look in the plane of the disc we see vast numbers of them, which to the ancients appeared like milk across the heavens (hence the name "Milky Way"). We live far from the centre of the disc, near to the edge. The earth is oriented such that those of us in the Northern hemisphere are forever forced to peer out towards the edge of the galaxy, whereas in the Southern hemisphere it is possible to look right into its heart. In the north, where modern city lights pollute the night sky, it is hard to discern the Milky Way, but in the desert of Australia or from the Andes, where the darkness is split by the full diameter of the galaxy, one can still experience the ancients' vision of milk spilled across the sky. Looking through the plane of the galaxy like this, into the "Milky Way," layer upon layer of stars compete for attention and there are too many for individual stars to be distinguished. By contrast, if one turns and looks out of the galactic plane, most of the sky is black but dotted with individual stars.

There are many other galaxies that are discs like our own and we can look at them from afar through telescopes. There are some shaped like ellipses, others with spiral arms like a catherine wheel, and some that are S-shaped like an elongated snake. These are far from the ideal sphere that gravity would seem to demand and are the result of billions of years of evolution as the individual stars tug one another in a kind of cosmic dance. With the aid of super-computers it is now possible to see how these shapes emerged. Initially, the computer is given the coordinates of a hundred thousand points, effectively stars, that are distributed symmetrically and spherically. The inverse square law of gravitational attraction (see page 13) is programmed in and the points are allowed to move in response to this force. As a result we can now do what no human has ever before done, namely watch the evolution of a galaxy of stars—albeit on a computer simulation.

In a real situation, a galaxy will not start off with a perfect distribution of points, uniformly spread through a sphere, equidistant from one another, and all with the same masses. The computer programme can test what happens when small asymmetries are introduced into the initial distribution. It turns out that even a small asymmetry can lead to large distortions as time elapses. After all, we are talking billions of years in the real case and there is plenty of time for a small disturbance to do its worst. As two stars head towards one another under their mutual gravitational attraction, they may miss one another, slow down, and return for another try. The effect will be that they start to rotate around one another. Now magnify this to a hundred thousand points in the computer, or in the real universe to billions of stars: the entire galaxy will rotate either clockwise or anticlockwise depending on the slight asymmetry that started the motion. A slight increase in density of stars in one region will lead to a clustering such that dense and less dense strands begin to form. The computer model shows how these lead to spiral arms, as observed in the real universe. The overall spherical symmetry is lost; the small perturbations lead to new patterns, but with symmetry nonetheless. The galaxies remain the same if you imagine them rotated in their plane about their centres; if the galaxy has a number, N, of spiral arms, it looks the same when rotated through one-Nth of 360 degrees. (For example, in Fig 2.3, the galaxy with two arms is symmetrical when rotated by 180 degrees). The original underlying spherical symmetry has become hidden; new and richer structures and patterns have emerged.

This is one example of how a disturbance can cause a "natural" symmetry to be lost or "hidden." In recent years scientists have increasingly come to recognize that the hiding of symmetry is very important in nature's scheme and may be the key to understanding how structure emerged from the original perfection of the Creation. A theme throughout this book, and its ultimate purpose, will be to introduce this relatively unfamiliar notion of hidden symmetry and its remarkable implications. More familiar to most of us is the appearance of symmetry due to manifest imbalances in the immediate environment. This is so familiar that we readily accept the asymmetric as "natural" and the symmetric form as unnatural. To illustrate this, let's come closer to home.

Watch a film of astronauts, weightless in space, attempting to drink. As they squirt liquid from their drink containers, perfectly spherical drops of juice hover uncannily in the cabin. This is quite unlike our experience on earth where raindrops fall to ground and

Fig. 2.3 A galaxy with symmetric spiral arms. © Anglo-Australian Observatory. Photograph by David Malin.

are fatter at the bottom than the top. Nonetheless, symmetry still remains in their shapes. We live on the massive earth whose gravity keeps our feet on the ground. The asymmetry here is that we are small and the earth is big and so it is we that dance to its tune. At any spot on the earth, gravity singles out which direction is "down." Objects that form in the presence of a large gravitational force will feel a pull in the down-up axis that is absent in the latitude and longitude of the surface. The symmetry of three-dimensional space that led naturally to spheres is destroyed, though a two-dimensional symmetry in latitude and longitude can remain.

As the sphere is the perfect symmetry of three dimensions, so is the circle for the two-dimensional plane. Gravity has singled out the vertical as special but east–west and north–south are indistinguishable to a raindrop falling in still air. So although raindrops are elongated and fat in their lower end, they are symmetrical, having circular cross-sections, in the plane of the earth. Of course, in a rainstorm it is likely that there is also a stiff wind blowing which will spoil even this symmetry. A gale from the west will deform the drop in the east–west axis, while leaving the north–south alone: the circular cross-sections in the ideal case will be distorted. The overall shape of a drop will be affected by the relative strength of the gravitational force pulling it down, the force of the wind on its western face, and the cohesive forces that hold the drop together. Symmetry can give a guide and in simple cases, as we have seen, can even give the answer, but to work out the shape in this most unsymmetrical case is probably best left to a computer.

Symmetry, spin, and bath-tubs

That the sphere is natural when there are no "spurious" influences at work can be seen by pretending for a moment that the sun had turned out to be shaped more like a rugby ball than the spherical soccer ball. The obvious question would be what caused it to have grown fatter in one dimension than the other two? This is not an idle question since in the case of the earth this is the nature of things, there being a noticeable squashing of the globe such that the diameter between the north and south poles is smaller than across the equatorial circle.

The origin of this asymmetry is similar to that of the raindrop. In both cases nature has singled out one direction as special while leaving the plane perpendicular to it symmetric. In the case of the raindrop it was gravity that singled this out. In the case of the squashed earth it is its rotation that is responsible. The globe is spinning around, once a day, the axis of rotation being the line connecting the poles which breaks the three-dimensional symmetry. As a centrifuge throws material outwards, so the rocks of the earth are thrust outwards from the axis of rotation, fattening the equator and thinning the poles.

The rotation of the earth gives rise to other asymmetries in human artefacts, as I have already mentioned in connection with my trip "down under." In the Northern hemisphere, the sun is to the south and moves across the sky from left to right as you face it, whereas when viewed from the Southern hemisphere, the sun is in the north and traverses the sky from right to left. Thus the shadows move hour by hour in opposite directions in the two hemispheres, which is the reason why sundials in Australia are mirror images of those in Austria.

However, contrary to a common shibboleth, the earth's rotation is not responsible for the asymmetric corkscrewing of water down the circular symmetric plug hole in your bath. The real cause is a subtle form of symmetry breaking known as "hidden symmetry" which we shall meet later on. It is the widespread unfamiliarity with this hiding of symmetry that has led to the spurious "explanation" and claims that the rotation is in opposite senses in the two hemispheres.

When I challenge people about this I am surprised that there is confusion about which way the water is supposed to rotate and that few seem even to have tested to see if it does as they believe it should. The fact that it "goes down the opposite way in Australia" is taken as gospel in the sure knowledge that the particular bathtub in Newcastle, North–East England, is unlikely ever to be taken to Newcastle, New South Wales to enable a test. However, I have twice experienced such a test and with mixed success.

The first was when I went to South Africa by plane from London. As we sat at Heathrow airport awaiting take-off, I realized that the wash-basin in the toilet was about to be transported all the way from north to south. I quickly locked myself in, with the plan of filling and draining the basin a few times to see if there was a tendency for the water to go one way more than the other, and then to repeat the exercise on arrival to see if anything had changed. I carefully filled it and let the water settle before pulling out the plug. The whole experiment was ruined by the sudden whoosh as the suction extracted the contents so fast that it was impossible to tell whether they rotated at all, let alone in which direction. So much for my experiment of long ago. I forgot about this until recently, when I was on safari in Kenya, where the effect was demonstrated to me.

On the first day of the safari we set off from Nairobi and headed north towards the equator. In the middle of the barren scrub this special line on the globe has been marked with a sudden presence of souvenir shops, photo-opportunities, and a charlatan with a bucket of water who, for a fee, will demonstrate that the water changes direction upon crossing the equator.

To perform this sensitive demonstration the man produced an amazing assortment of artefacts. The water bottle was a rusty tin can from an old jeep. This was placed at the equator and used to fill a plastic cylinder that in a former life had been the canister for disinfectant spray. It was some 30 centimetres high and 10 centimetres in diameter and in its base there was a small hole which he covered with his finger to keep the water in. He filled the cylinder and floated two matchsticks on the surface so that we could see which way the water rotated when he removed his finger from the hole. A plastic washing-up bowl to catch the water completed the apparatus. Thus prepared he started his patter.

We were informed that it was necessary to go 20 metres north of the equator, to where he had placed the wash-bowl to catch the water after it had drained out, and then 20 metres south where the effect of the earth's rotation would be felt. We duly followed him. He blinded us with science explaining that the water should rotate out clockwise in the north but anticlockwise in the south, and then began the demonstration. He removed his finger from the hole in the base and the water drained out. He tilted the can so we could all see that the matchsticks were indeed rotating clockwise as the water poured out. Being both a good conversationalist and conservationist he continued to talk while he poured the water from the wash-bowl back into the can and we all marched solemnly the short distance southwards, crossing the equator, until we reached the symmetric point 20 metres into the Southern hemisphere where he prepared to repeat the show. All the while as we talked he was continually jabbering and juggling with the matchsticks and water. Precisely how he managed to do the trick—as the water did miraculously rotate in the opposite direction following a mere 40-metre stroll down the singularly unimpressive dusty road—I could not tell. I asked him "How did you do that?" meaning "What was the secret to the trick?" Misunderstanding the reason for my question he looked at me seriously and with awe in his voice, as if announcing the second coming, exclaimed "It's physics!" and, as way of further revelation, uttered in a conspiratorial hush, "Coriolis effect."

I was amazed that in the middle of the African highlands, innocent vacationers now receive physics lectures. The theory was excellent and I handed over my shillings in amusement, but the demonstration was sheer quackery. To reveal why, I must explain what is going on with this water twister.

The secret is similar to the way that an ice-skater, spinning like a top, with their arms outstretched, suddenly rotates faster upon pulling their arms inwards. The natural law responsible is that the angular momentum of a freely orbiting body remains the same, or, to put it another way, if the radius of its orbit gets smaller, its speed correspondingly grows. So when the skater's arms are outstretched, his or her hands are moving around a larger circle than when they are clasped in to the waist. To preserve the angular momentum the skater spontaneously rotates faster. This is forced on the skater by the laws of nature; it is unavoidable.

Now imagine yourself at the equator as the earth spins around its axis. In 24 hours you will go around once and have been transported through 24,000 miles of space while remaining fixed on the surface. If you had instead been shivering at the research station near the South Pole, you would have spun nearer to the axis and travelled around a circle of only a few miles. Your angular momentum would be much less in this case than on the equator. If you had started on the equator and moved towards the pole, you would have moved onto ever smaller circles of rotation and, as when the ice-skater's arms moved inwards, you would have sped up.

Move north or south from the equator and you start to overtake the rotating earth as its surface swings east towards the rising sun: you move eastwards faster. Heading north and east is a clockwise motion whereas heading south and east is anticlockwise. If you are part of the surface, like a mountain or a river-bank, there will be stresses and strains from this but you will stay fixed, whereas free-flowing things, such as the atmosphere or water, will be shifted relative to the surface. This is the source of the great circulations in the atmosphere that generate our weather. In the oceans it causes some of the major currents; in rivers, over the aeons, it may cause one bank to erode more than another. However, when water starts to drain down the plug hole in your bath, the distance involved is so small that this Coriolis effect (named after the nineteenth-century French engineer and mathematician Gustave-Gaspard Coriolis) is nugatory.

Nonetheless there have been careful experiments attempting to detect it. One, which was reported in the journal *Nature* many years ago was performed in, of all places, Stillwater (!), USA. There are several things that will override the subtle Coriolis effect in

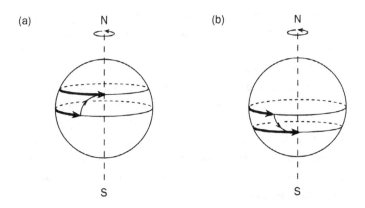

Fig. 2.4 The Coriolis effect: (a) Moving from the equator, northwards to the Tropic of Cancer: the conservation of angular momentum makes the object move faster than the surface of the earth. This north to east movement is a clockwise curve. (b) Moving southwards, towards the Tropic of Capricorn, has a similar outcome. The diagram is a mirror of the Northern hemisphere one; the movement is now anticlockwise. (c) The Coriolis effect causes the large scale circulations in the atmosphere as in this time-lapse exposure satellite image of Hurricane Andrew, NASA/Goddard Space Flight Center/Science Photo Library.

normal circumstances. In order to have any chance of seeing it, first you would need to have a symmetric arrangement where the plug hole is circular and central in a large basin. Second, you must ensure that after filling the basin, many hours have elapsed so that all motion in the water has ceased, as the slightest disturbance is sufficient to distort the symmetry. Third, air currents above the water can disturb it enough to spoil the experiment and so you need to cover the basin. Finally, the basin should be such that the plug can be removed from beneath without disturbing the liquid above and setting up unwanted vortices. After doing all of this you might have a chance of detecting the Coriolis effect.

In the actual experiment in Stillwater, all these conditions and more were met and the result was as expected: the water drained out in the direction that Coriolis would have predicted. The authors noted, however, that this could still have been due to some other overlooked asymmetry in the design of the bath and that to do a definitive test would require them to take the same bath to the Southern hemisphere and then repeat the whole exercise. This they never did and the only example known to me where an "experiment" has been carried out with the same apparatus in both hemispheres is that of the entrepreneur in Kenya.

Being aware of how subtle the real effect is, you now see how it is impossible for a definitive demonstration to be made while walking down a windswept dusty street in the Kenyan highlands. Apart from the obvious vibrations, and the fact that 20 metres north or south of the equator is far too small a distance to become sensitive to the curvature of the earth, there was another, more important reason why it could never be done there. The critical feature underpinning the effect is that on crossing the equator you are at the furthest point from the axis of the earth's rotation such that whether you head north or south it is as if the ice-skater's arms are being pulled inwards and so you will start to rotate faster than the solid earth. However, the road was on a slope and so, as we walked the 40 metres from north to south, we were going downhill and always getting nearer to the earth's axis.

Beyond the sphere

At the moment of conception, life begins as a spherical single-cell embryo, but nine months later this perfect spherical symmetry has been utterly transformed. While it may be hard to tell one baby from another, it is easy to tell a baby from a sphere. How did this remarkable breaking of symmetry come about? This highly structured being with a head at the top, legs at the foot, and asymmetric within from left to right demands an explanation.

The spherical sun and near spherical earth are the result of gravity pulling equally from all directions. However, gravity does not have much to do with the human form; we experience gravity through our attraction to the earth but apart from this it has no immediate effects for us. As all solid matter, we are held together by other forces—electric and magnetic. The force of gravity made Newton's apple fall to earth. What stopped it carrying on all the way to the centre was the solid floor which is there courtesy of electric and magnetic forces. These forces (now referred to collectively as the electromagnetic force) also obey an inverse square law, diminishing as the square of the distance, as does gravity, but intrinsically much stronger. (The magnetic field of the earth, for example, with the magnetic poles thousands of kilometres apart, can nonetheless swing a compass needle around.) They are also more diverse than gravity in that whereas gravity attracts every bit and piece in proportion to its mass, electric and magnetic forces can attract or repel (a familiar example being the powerful attraction when the north and south poles of two strong magnets are joined, as against the repulsion felt when two similar poles approach one another). This leads to more variety in structure, with its own symmetries, and rarely anything so simple as a sphere.

These electric and magnetic forces can raise mountains and support the walls of canyons. Yet, compared to the size of the world as a whole, these are mere surface irregularities, their sizes trifling relative to that of the earth. Humans are highly structured and we might wonder whether in principle the mountains and valleys on the earth's surface could have been so grand that elongated limb-like features could have distorted the shape of our planet to be nothing like a sphere. However, mountains on the earth cannot be more than about 10 kilometres high. If they were to be forced higher by the tectonic motions of the earth, their weight would break the electrical bonds at their base. The effect would be similar to the way that a skater crosses the ice. The skater's weight puts great pressure on the knife-edged skates which in turn causes the ice to melt. The skater skims across on a thin plane of water which freezes again as he or she passes. A similar thing would happen with a huge mountain. Once above 10 kilometres high, its weight would exert so much pressure at the base that the rocks would melt. Below this height the rocks can support it.

The maximum deviation from a "smooth" surface for the earth is therefore only 10 kilometres relative to a radius of some 6000 kilometres—much less than one per cent. That is the reason why the solid earth has remained so faithful to the underlying spherical symmetry of gravity. What if the earth had been much smaller? The amount of material would be much less, so its mass and the effects of its own gravity would be correspondingly reduced. The strength of the electric and magnetic forces within its rocks would however remain the same. The spherical influence of gravity would become finally overwhelmed by the effects of these electromagnetic forces were a planet made of solid rock smaller than about 500 kilometres across. Humans are so much smaller than this that, for each individual, gravity is inconsequential. It is electromagnetic forces that bind atoms and molecules tight together giving shape and form to solids. It is electromagnetic forces that determine our shape and that of the emergent embryo.

It is arguable whether beauty is only skin deep but the embryonic spherical symmetry certainly goes no deeper than this. The newly formed embryo is smaller than the eye can see but it is nonetheless huge on the scale of the molecules that are contained within its spherical skin. Seen under a microscope that nascent embryo is revealed as far from a perfect sphere; it has a complex

internal molecular make-up, consisting of rich patterns of inter-
twining carbon, nitrogen, and oxygen atoms in helical chains of
DNA. It is these templates that will eventually signal the cell to
differentiate the three dimensions of head to toe, front to back,
and left to right. The problem though is: why and how does this
happen? Evolution explains some of the why: our head held high
has obvious advantages, while the hard spine defending the rear
and eyes peering forward can also be understood. However, the
differentiation between left and right (such as the heart looping to
the left, the stomach to the right) is still a mystery.

What causes this and when it happens in the foetal development
are questions that fascinate embryologists. An answer could lead
to ways of eradicating genetic defects. This promises to be a novel
application of a natural asymmetry.

Chapter 3

Through the looking glass

In the film *Being There*, Peter Sellers plays the role of Chauncey Gardner, a simple man who has never ventured outside the house where he tends the garden of a recluse. Chauncey's only experience of the outside world consists of what he has seen on television, and watching television is his main pastime. The owner of the house dies and Chauncey sets off on his adventures into the wide world of Washington, DC, taking with him one change of clothes, an umbrella, and a remote control for the TV.

In one scene Chauncey finds himself in front of a shop window where several televisions are on display. After flicking the channels with his remote control he inadvertently triggers the security cameras; all the screens go blank except for a large one that dominates the display on which appears a life-size image of Chauncey. Knowing nothing about closed-circuit television he assumes that he is looking in a mirror and is seeing his reflection. He starts to walk off to the right and then stops abruptly: his image on the screen has moved off in the opposite direction. He tries moving the other way, and his image also turns about and refuses to keep track with him. He waves his umbrella in the air and the Chauncey on the screen raises its umbrella, not by the arm immediately facing Chauncey's raised hand, but by that on the other side.

He is so astonished that he steps back into the road and is hit by a car. This ironically is the beginning of his good fortune—but that is another story. It is his reaction to seeing his image doing all

KOSMO by Kirschen

THE MIRROR DID **NOT** SEEM TO BE OPERATING PROPERLY.

©1998 Kirschen

Fig. 3.1 "The mirror did not seem to be operating properly."

the "wrong" things in what he perceived as the mirror that is so amusing because it appears superficially to be naive, while it is in fact profound.

Our ability to distinguish left from right is a sophisticated one. When a picture is reversed, few people notice. There is a simple test that you can do which will illustrate this. In many countries stamps and coins are embellished with a profile of the president or monarch. In Britain the Queen's profile is shown: without looking, can you say whether she faces to the left or to the right on stamps? Does she face the same or the opposite direction on stamps and on coins? These are objects that one sees hundreds of times a day and yet if you ask a large group of people the questions, nearly as many will get the answers wrong as correct.

Distinguishing left and right is more of a reflex action than a conscious one. Many young children tend to reverse letters and words, confusing their "p"s and "q"s, or "d"s and "b"s; while most children soon learn to distinguish these mirror images, dyslexic children continue to confuse them.

We are much less aware of the distinction between left and right than we might at first sight admit and it is this "stupidity" that is at the root of much of the fascination with mirrors. Lewis Carroll wrote an entire children's story based on this in "Alice Through The Looking Glass." When Alice enters the looking-glass room she spies what appears to be an ordinary book, except that its script seems to be written in some strange code. She realizes that it is a "looking-glass book" and holds it up to the glass so that "the words all go the same way again." Alice knows, as we all quickly learn, that the reflection of a reflection looks the same as the original. But why is it that a mirror inverts left and right and not top and bottom? It is obvious from experience that it is so, yet a flat, square mirror should seemingly care naught whether you turn it through 90 degrees: why did it invert your left and right and then, after being turned through 90 degrees, continue to select left and right along with what was previously the up and down axis?

I will answer this later in the chapter. The solution is easy once one gets some misconceptions about symmetry sorted out. To prepare ourselves, we will begin by looking at the best-known example of mirror images: ourselves.

The human body

Although people are symmetrical neither from front to back nor from head to toe, the human body from the outside appears to be broadly symmetric left to right. Apart from self-selected cosmetic asymmetry, such as the parting of one's hair or the greater muscle development on the right half of the natural right-hander, there is only one externally visible "natural" asymmetry and that is in the tendency of a man's left testicle to hang lower than his right. This is clearly a case where symmetry is not desirable as two equally "slung" testes would keep bumping into one another. In addition to the discomfort this would cause, this would be detrimental to the healthy generation of sperm. Here is a case where it is plausible that natural selection has led to the asymmetry—though why it should have chosen to lower the left rather than the right is part of the mystery.

Apart from these and other minor exceptions such as scars, vaccination marks, and freckles, humans superficially appear to be symmetrical. However, such symmetry is only skin deep. In all vertebrates, the heart and stomach are on the left, the liver and appendix on the right, though one person in ten thousand has their heart in the wrong place (namely the right instead of the left). If the heart is on the "wrong" side, all the other organs tend to be also which can give surgeons a nasty surprise if they are not well prepared. The heart loops to the right (in "normal" people) while the stomach loops to the left. In those rare individuals where the organs are in their mirror position relative to the norm, the directions of the loops are also inverted. Whatever determines the orientation for one organ, determines it for them all. The mirror inversion is absolute. Somehow the original microscopic spherical embryo becomes chiral, left–right asymmetric; the whole then choosing either the "standard" or the mirror version.

Our right lungs are larger than our left, while the left half of our brain is slightly larger than the right. Asymmetries in our brain and in the way that our nervous system is connected control much of our behavioural asymmetries. The nerves that connect our brain to our muscles cross over so that the left side of the brain controls the right side of our body and vice versa. The right-handed dominance is therefore equivalent to a greater control by the left half of the brain. The left side of the brain is concerned with language, reading, writing, speaking, and hearing, while the right controls our visual, artistic, and musical awareness. Thus, if a stroke damages the left side of the brain, speech impediment often occurs.

Our right and left eyes are also controlled by our asymmetric brain such that we are all either left-eyed or right-eyed. You can test which you are by pointing at an object that is about two to three metres away, then cover the left eye and see if your finger is still pointing at the object or is directed slightly away from it. Then repeat the exercise by covering the right eye rather than the left. If you finger was still pointing at the object when your left (right) eye was covered then you are right (left)-eye dominant.

Left- and right-handers

One of the most familiar examples of mirror symmetry is the fact that humans use one hand or the other, but not both, for skilled tasks such as writing. Are you left-handed? If so, you will be all too aware of the way that society discriminates against you. Left-handedness is not that uncommon—one in ten people on average are—and yet we are surrounded by artefacts designed with right-handers in mind. For example, it is easier for one's right hand to twist clockwise than anticlockwise, the left hand showing the opposite preference. In years gone by ropes were made by humans physically twisting hemp. When right-handed people did the twisting the result was a rope with a left-handed twist, left-handers producing a right-handed thread. By looking at old ropes you can tell the handedness of whoever twisted it.

Screws, being helical, can be right-handed or left-handed. Mass-produced screws are usually right-handed which means that they move forwards when turned in the clockwise direction, the natural motion for a right-hander. There are occasions where asymmetry is necessary in order for artefacts to work efficiently. The inherent asymmetry of the left- and right-handed screws has important practical uses; for example, the axles of Land Rovers are screwed in opposite directions at each end so that they do not unscrew accidentally as they bounce over rough terrain, bicycle pedals also have this feature. The design of screwthreads, or corkscrews, ensures that it is easier for right-handers to insert screws or to uncork wine.

Mass-produced scissors are made for right-handers and when used normally the two blades come together and cut. However, if you put them in your left hand, the normal motion will force the blades apart. For a left-hander to use scissors it is necessary to grip the handle with the thumb and then pull it back to make the cutting action. Try to cut the fingernails on your right hand by holding the scissors in your left and you will understand what I mean. This is very awkward and short of buying "left-handed" scissors from a specialist shop, it is often easier for a left-hander to use their right hand for "low-skill" tasks. This "if you can't beat 'em, join 'em" attitude plays havoc with simple attempts to determine the incidence of left-handedness.

What is a "left-hander"? This is not so obvious as it seems. Which hand do you use for writing, sewing, cutting with a knife, spreading butter, hammering, throwing, and a range of other handed tasks? Is a right-hander only a person who answers "right" to all of these? If you regard yourself as a left-hander, do you perform all of these tasks with your left or do you mix and match? If so, is it because you have given up fighting the world of right-handed artefacts and decided to join the bandwagon whenever you can? In an early investigation into handedness, it was deemed that the subject was right-handed if they performed even a single task with their right hand, notwithstanding that they may have done all of the others with their left. Not surprisingly this study resulted in a relatively low proportion of left-handers.

Many tasks involve the use of both hands, but even here a left–right asymmetry emerges as one hand takes the more technically demanding role. For example, pianos are normally made such that the essential melody in the higher registers is where the right hand rules. When playing the violin "right-handed" it is the bowing arm that is the right, while the left fingers depress the strings to form the notes. The guitar is similar; Paul McCartney plucks the strings with his left hand and forms the chords with his right and as such is a rarity. The higher skill seems to be in using the dominant hand to bow or pluck rather than to perform the more passive action of depressing the strings. While one might dust a table with either hand, there is a clearer preference for which hand you hold a dish in when cleaning it.

It is an unusual individual who cannot learn to use the non-preferred hand for all tasks other than perhaps writing. Most people, when asked, will identify themselves as right- or left-handed based on their writing hand. Moreover, we identify others accordingly. Writing is a highly technical activity and the least likely of all tasks to be performed equally well by either hand—even by individuals who happily are ambidexterous in more mundane tasks. If handwriting defines handedness, then about 10 per cent of people are left-handed.

In almost all animals that have been tested, even including the great apes, there is for each individual a preference to use one hand rather than the other. For the monkey or ape species, as many are left-handed as right-handed. However, there is some controversy here: whereas gorillas certainly seem to be 50:50, left or right, it is possible that for chimpanzees, 60 per cent are right-handed.

For humans, it is very different—nine out of ten are right-handers. This dominance appears to be standard throughout the world and goes back as far as we have been able to trace throughout history. There is a strong and persistent historical and cultural heritage where left-handers were regarded as "sinister" (from the Latin word for "left"), vilified, and even persecuted. Culturally, the world is biased for the right: the place of honour is on the right hand of the host; Jesus ascended into heaven to sit on the right hand of God (according to Christian tradition); it is God's right forefinger that creates Adam in Michaelangelo's famous painting. Dexterity, adroitness, and rectitude are all good things; by contrast, "lefties" are gauche. If left is sinister, then right is right!

The angles of the lines in Leonardo da Vinci's cartoons show that he was left-handed: the diagonal from top left to bottom right is the natural sweep of the left-hander, as in Leonardo's drawings, whereas right-handers naturally draw along the other diagonal. As an artist he was exceptional, and as a left-handed one he was the exception (as the signature of right-handed artists dominate in quantity if not in quality). Even ancient cave drawings show, by the angles of the lines, that the artists were right-handed. The grooves in ancient stone carvings, or the cutting of flints, show a right-handed dominance in prehistoric peoples.

Teeth of early humans from the Pleistocene period, half a million years ago, also show right-handers dominating. Having hunted and killed animals, these cave dwellers would grip the meat between their teeth and cut it with sharp flints. Often this also cut grooves across their teeth, many of which from this period are marked with scratches that run from the top left near the gums to the bottom right. In all of these creations, through to modern times, we see this tell-tale signature in the natural direction for right-handers.

There have even been attempts to quantify the amount, historically, of left versus right, to see how it compares to modern times. One example is in the Bible where, in Judges, there is mention of the Benjamite army that had a special group of 700 slingshotters who were all left-handed. The army numbered 27,000 men in total, and so the 700 left-handers works out at 1 in every 40—some 2.5 per cent. This is lower than the one in nine that modern measurements find but this is perhaps not surprising as in the case of the Benjamite army we are comparing the number of left-handers with a special skill, namely slingshooters, with an entire army. Consequently, the 1 in 40 is certainly an underestimate because there was nothing to imply that this special group contained all of the left-handers in the army; there were surely some left-handers who were no good at slingshooting.

The asymmetric brain

What causes left- or right-handedness? There are clues that it is all in the mind (in the neurological rather than the psychological meaning of the phrase). I mentioned earlier that our brains are left–right asymmetric in that they control respectively our right and left sides. There is asymmetry in function too: our brains are specialists. The brain's left hemisphere specializes in verbal abilities and is the dominant hemisphere for humans. The left brain

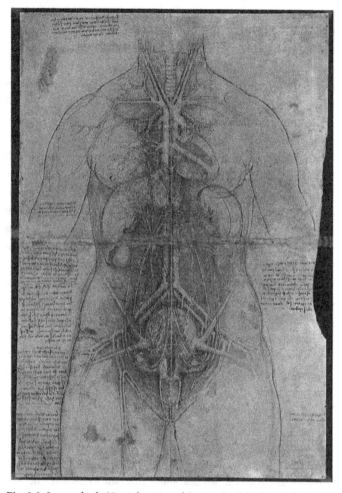

Fig. 3.2 Leonardo da Vinci drawing of the trunk of the female body. The direction of the shading from upper left to lower right is the natural one for a left-hander. This drawing also illustrates the left–right asymmetry of the body's internal organs. The Royal Collection © Her Majesty Queen Elizabeth II.

also controls the motor mechanisms of the right side of our body and so when writing with the right hand we are using the same left-brain that specializes in language. Hence one half of our brain is involved in the two tasks, thereby releasing the other half for other activities. Using the same hemisphere for both tasks cuts down on information processing. The human brain and nervous system has thus developed in an energy-efficient way.

One theory is that the specialization of the brain hemispheres in humans has been driven by the development of language, which is unique to us, and that in turn this has produced the uniquely human handedness. It is still the case in some cultures that when a child is born naturally left-handed, the parents will encourage it to use its right hand. When I was at school around 1960, there was a boy in my class who stammered, this supposedly having been brought on by him being forced to write right-handed. Whether that was really true I do not know (and bear in mind that young children enjoy creating fables) but it was certainly the case that he grew out of it when he reverted happily to his natural left-handed script. More serious studies suggest that there is indeed a correlation whereby some left-handers, forced to write with their right, develop speech impediments, at least in childhood. The fact that language and writing are both controlled by the same half of the brain could explain why these learning and behavioural problems arise and also why they tend to disappear once the individual is encouraged to revert to their natural left hand.

Although the fundamental cause of stuttering is unknown, some recent medical studies show that stutterers use the two halves of the brain in a different way to those with normal speech. It is possible to look inside the living brain by means of positron emission tomography (PET) scans (see page 209). This has shown that when a normal person speaks there is a single centre of activity in the left half of the brain, whereas for stutterers both halves of the brain are activated. In order to speak efficiently, presumably these two centres have to be coordinated, which is difficult across the corpus callosum (the neural fibres that connect the two halves of the brain). If these discoveries are the explanation, then it shows that left–right asymmetries are involved in stuttering, but the connection with enforced dexterity for left-handers is at most a secondary effect.

Society favours right-handers, but scientific studies have shown that education and environment are not the explanation for right-handers winning the lottery of life. When children are adopted, and thereby freed from any influence of their natural parents, they continue to maintain the same ratio of right- and left-handers, irrespective of the handedness of their adoptive parents. This suggests that an "intrinsic" hand preference has emerged very early in life. So if our choice is not altered by our upbringing, is it influenced by our genes?

There is some evidence that handedness, and related asymmetries, do run in families. Take the right-handed and left-brain correlation. It is known that damage to the left brain, by stroke, accident, or at birth, can affect the motor mechanisms of the right side dramatically. However, right-handers who have left-handers among their near relatives seem to recover better from such injuries. This supports a link between familial handedness and brain organization.

Handedness and genes

There is also a tendency for left-handers to be more likely to have a left-handed parent, sibling, or child than right-handers. This supports the idea of a genetic effect, but there are counterarguments favouring social pressures; for example, a left-handed child with a left-handed parent will stay left-handed, whereas with a right-handed parent it may pick up right-handed cues socially, at an early stage. There will indeed be some social pressures but overall there does appear to be a genetic effect underlying handedness. The best evidence that genes are important may be that children of two right-handed parents are more likely to be left-handed if one grandparent is left-handed. It is plausible that the asymmetries of the brain and handedness are linked and that energy efficiency, amplified by evolution over the generations, underpins it. But how is this information passed on? Is a gene responsible?

A favourite way to test for inherited genetic effects is to study monozygotic twins that have developed from the same egg. Monozygotic twins are colloquially known as "identical" twins, though this is a misnomer in the case of handedness because one in five pairs of monozygotic twins have opposite handedness. This is similar to the ratio found in the more common dizygotic twins and hardly different from the population at large (left-handed pairings for the twins being marginally more than chance). If handedness is primarily genetically controlled then one might expect that both twins in the monozygotic pair should have the same handedness. However there are two problems, at least, with this. First, genetics does not imply that monozygotic twins should be truly identical, only that they are more similar than dizygotic twins. A second problem is the mirror-image effect, where one of the pair has physical characteristics that are the reverse of the other. For example, if one twin has its hair whorl clockwise, the other is anticlockwise. There is also the tendency for young twins at play to sit face to face and to imitate one another. To mimic a right-handed twin, the other twin must act in a left-handed manner. Whether or not this persists into writing remains unclear but this uncertainty does make it difficult to draw firm conclusions from what at first sight appears to be an ideal situation.

The balance of opinion is that handedness does appear to be inherited but the selection mechanism is complicated. More men are left-handed than women and this has led some researchers to suspect that if there is a gene that determines handedness, it might be located on the X and Y chromosomes that determine gender. However, opponents point out that there are many characteristics that differentiate between the sexes but are due to genes that are not on the sex chromosome. The search goes on.

If indeed handedness is an outward signature for asymmetry in the brain, then the discovery of the gene (if that is indeed the cause) could open up new areas of brain research. These could include how people differ in brain structures and in what ways this affects personality and language structure. Creativity is a right-brain activity and the gene might also give clues to this gift of genius. Whether this will be reality or merely the hyperbole of over-excited writers of science fiction is for the future. Certainly it will be a long and helically twisting road between identifying the gene and designing "Einstein to order."

Tenez la droite

Traffic provides an example of the advantages of mirror asymmetry and evolution. Motorists could travel on either the left (as in Japan or Great Britain and parts of its old Empire) or on the right (as in most other countries). Safety demands that the symmetry is broken so that one side or the other is adopted in a local region. The reasons why different countries have ended up with the left or the right as their choice can be traced to their different histories.

A preference for the left-hand side emerged in the days when riding horses was the norm. Historically, before stirrups came on the scene, the strong right arm of the rider was the key. You stand astride the horse facing forward at the roadside. Your stronger (right) arm hauls you on and so you must stand to the left of the horse. Having mounted, you then proceed in the direction that you are facing—hence on the left-hand side of the road. With stirrups the result is the same as you grasp the reins in the left hand and the pommel in the right, if you are a natural right-hander. Then with left foot in the stirrup, you tug with the right hand and swing into the saddle. As before, you will ride away on the left-hand side.

Once on the horse, this left-hand drive helped self-defence. Your sword being on your left, you pull it out with your right hand and defend yourself from passers-by on your right. The horseless peasants, meanwhile, walked facing the traffic, on the right—a practice still followed for safety by pedestrians on country lanes today.

The French revolution enforced right-hand travel, supposedly in sympathy with the peasants and contrary to the horse-riding establishment. Napoleon (who was left-handed!) adopted this and spread right-hand travel around his empire.

Napoleon's influence led to some bizarre consequences. Those parts of Austria that he had conquered duly adopted right-hand drive, while the rest kept to the left and remained this way until the First World War. In Italy, cities whose politics were sympathetic to the revolution also drove on the right, while the more conservative drove on the left. Milan and Turin drivers kept on the left in town and then changed to the right once outside—a practice that remained until May 1926; thankfully traffic was light in those days.

Quality cars from France and Italy were designed with the steering wheel on the right up to the 1950s. This was because the drivers were likely to drive over the Alps and it was thought to be safer to position the driver near to the drop. Some transalpine trucks were designed this way until recently, though it is less common today since better roads and safety barriers have reduced the risks.

Historically, when travel was rare, regions of left- or right-hand preference happily coexisted. Today however, large land masses, connected by multiple roads, demand that the same traffic flow direction be adopted over entire subcontinents. Thus islands such as Japan or Great Britain can maintain left-hand travel, as can the island continent of Australia and India (separated from the rest of Asia by mountain ranges). Indonesia was a Dutch colony and as such a natural for right-hand travel; however, the first car was that of the Sultan who bought a British Rolls Royce, since then left-hand travel has ruled. The USA, Canada, and mainland Europe all drive on the right. Until 1967 Sweden drove on the left, by which time increasing road links with its right-hand neighbours, combined with the economic advantages of building cars for export to the dominant European and American markets, led to them changing to the right-hand side. In this modern example we can see an economic "Darwinian selection" where an initial random asymmetry, with some countries choosing the right and others the left, becomes reinforced in one direction such that a (near) global asymmetry can result.

Asymmetry and evolutionary biology

Left–right asymmetry is more profound in humans than simply a matter of which hand we prefer. We have a tendency to impose symmetry onto images that are not symmetrical. Symmetry equates with good genes while asymmetry, whether intrinsic or caused by damages from conflicts, implies problems.

The faces on statues in the Tuileries or in some artworks are mirror symmetric and it is this symmetry that creates their eerie beauty. For humans it is very different. We are stuck with asymmetric faces and it has been shown that we seem to prefer asymmetry when seeking our mates. According to some evolutionary biologists, this has become reinforced through the generations such that there are characteristic and measurable asymmetries visible in all our faces.

It is extremely rare for someone's nose to be perfectly vertical, their lips or eyes to be horizontal, their ears to be the same, and so forth. It was only when I was fitted with reading glasses that I discovered that for forty years I had been unaware that my left ear is nearly a centimetre further back than my right. A bizarre illustration of this can be achieved by taking a full-face photograph of yourself and also of your mirror image. Then cut these images vertically down their middle. Finally join the left half of the normal image to the right half from the mirror to make one complete face, and do the corresponding combination with the right normal and left mirror halves. When the resulting images are compared with the original, it is as if one is looking at images of three completely different people.

Dahlia Zaidel is a cognitive neuropsychologist who has used this asymmetry in a novel way with unexpected results. First she photographed hundreds of full-face portraits—typical passport style with a dour stare such that if you look like your image you really need the vacation. Then she divided the facial images vertically into left and right halves, and printed mirror images of each. With the left and right original halves and their mirror images she could create four faces: the original and its mirror image but also two symmetric versions that had never before been seen (a left half and its mirror image, and a right half and its mirror).

The novel versions are symmetric, whereas the original and its mirror images are asymmetric. The question that she posed was how people would react when shown the faces. If they were not told which were real and which artificial, what would their preferences be? First she showed images of mens' faces to women. There was a strong tendency for the women to prefer the asymmetric originals (or their mirror images) over the symmetric artificial composites; the two varieties of composites were rated equally. When men were shown images of women they too preferred the natural asymmetric faces over the artificial ones, but here there was a surprise—of the artificial versions there was a clear preference for the ones where the right-hand side had been married to its mirror image, relative to those where the left-hand side had been married to its mirror. Somehow the asymmetry in womens' faces is not random and is such that males prefer the right half over the left. Dahlia Zaidel associates this with the left–right asymmetry in the way the brain processes information. To see how we need a digression.

I have mentioned already how the left half of the brain specializes in communication and the right is more visually responsive. These tendencies also include that of the left side to specialize in incongruity, while the right seeks order and beauty.

The way this was demonstrated was as follows. First it involved placing a screen in front of the nose so that the left and right visual hemispheres are separated. Then the researcher placed images in one of these hemispheres; if in the left hemisphere, it will be processed by the right brain, while if it is in the right hemisphere, the other half of the brain takes over. The researcher flashed many

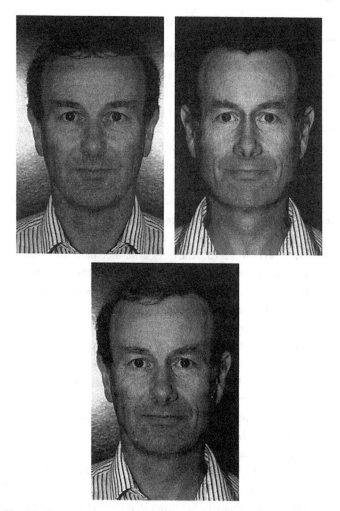

Fig. 3.3 Two symmetric and one asymmetric versions of the author.

images quickly and then asked the viewer to recall them. These images could be "normal" like an eskimo and an igloo or a doctor at work or a Rembrandt; alternatively, they could be incongruous, such as an eskimo and a giraffe or a doctor injecting into a coffee cup (yes, such images as these were actually used!) or paintings by

Salvador Dali. When shown in the hemisphere that is processed by the left side of the brain, the viewer recalled the incongruous images better but when shown in the other hemisphere, it was the "normal" ones that tended to be remembered.

So how does this affect one's response to images of faces? When a person is facing you, or when you look at a full-face photograph, their right cheek, ear, and eye are in the left-hand hemisphere of your vision and are processed by the right side of your brain. Succinctly: their right half is processed by your right brain, and their left by your left.

Now, it is the right side of your brain that responds to beauty and the left to incongruity, so a face which is beautiful on its right half and less so on its left, will be preferred—at least by males apparently. So how is it that the faces of women have miraculously managed to make themselves more beautiful on the right than on the left—for that is the implication of Dahlia Zaidel's results. It is at this point that evolutionary biology enters the story.

The claim is that men have been primarily attracted by a pretty face and chosen their mates this way from prehistoric times. Faces have always been asymmetric and those females whose right half was pretty would be favourably noticed by the beauty-seeking right brain of the Stone Age male. Eve and her sister Vev might have had faces that were mirror images of each other, but if it was Eve that had the ugly left side and pretty right side, she would attract the male eye; her sister Vev, with pretty left but ugly right, would not stimulate the asymmetric male brain. The result would be that Eve would be a more favourable mate than Vev and the accidental tendency in her genes that had created this advantage would be passed on and reinforced through the generations. Modern women, according to this theory, have evolved with their beauty on the right as a result.

The explanation fits the facts but I am not totally convinced that the phenomenon would have been predicted had it not first been observed. Prediction is the real test of a theory. After all, economists are very successful at explaining why the stock market rose sharply yesterday but few of them appear to have become million-aires by anticipating the move. Nonetheless, it is interesting that there are asymmetries in faces as well as in the brain and that they appear to correlate in this way.

Whether evolutionary biology and asymmetry are related is still controversial but it is clear that the brain is asymmetric, sharing out its tasks. This is not really surprising. The brain uses more power than anything else in the body. To have the entire brain involved in processing each and every piece of sensory input would be exhausting and inefficient. Electrical currents flash around the nervous system, exciting the brain which processes vast amounts of data every fraction of a second—faster than it takes me to write this or you to read it. Minimizing the route that the information has to travel would confer advantage and so it is preferable that different parts of the brain specialize in different tasks. This would lead to an asymmetric brain. How far this is responsible, through evolutionary biology, for other asymmetries in our make-up is still open to debate. In any event this cannot be the whole story as it alone cannot explain how humans are the only animal with such major functional asymmetries. We would expect that all animals would have had the same desire to reduce energy consumption in the brain and nervous system.

Animals and insects

There are many examples of symmetry and asymmetry among creatures. Some of these are the results of accident or chance, while others have evolutionary advantage.

Some animals have horns with helical twists. A helix, like a spiral staircase, can twist up and to the right or to the left. Antelopes and rams have a left-handed helix on their left horn and a right-handed one neatly balancing it on the right horn. This matching is known as "enantimorphism." Not all helices in the animal kingdom are like this, however. The arctic Narwhal whale is about 5 metres long; the oddity is that its left tooth grows straight forward to form a 2-metre-long ivory tusk with a pronounced leftward spiral. There are no known examples of a right-handed Narwhal.

Contrast this with the world of birds. In Europe there is a breed of finch with an asymmetric beak whose upper mandible points to the bird's right and the lower points to the left. This asymmetry is so striking that it has given the "crossbill" its name. Why is the bill of the crossbill crossed? There are two ways the bird could eat the seeds it needs as food: it can prise open a pine cone and get the seed into its open mouth; or it can insert a crossed bill and screw out the seed directly into its closed mouth. (The distinguished physicist, Denys Wilkinson, once commented that the muscles to close a mouth are tougher than those to open it—as is well-known to all who have sat on committees.)

The emergence of the crossbill therefore owes something to evolution, as does the Narwal's toothy tusk. In Europe all crossbills have their upper mandible heading right, whereas in North America the asymmetry goes the other way, with the upper bill pointing left. Nature does not care which way the symmetry is broken as long as there is advantage in it. Then over the generations, the "accidental" dominance of one or the other becomes encoded and established. Presumably the crossbills of America and Europe developed independently and formed two separate "domains" of asymmetry, whereas Narwhals are so closely located that effectively only one form developed. That it was left rather than right was chance. How such chance happens will be dealt with later.

The turbot, flounder, and other flat fish are flat because they swim on the bottom of the ocean and it improves their chance of survival. Fish swimming in the seas are the prey of larger predators; they can be attacked from above or below. Swimming at the seabed eliminates attack from below and this is where the story of the flat-fish begins. Swimming on the bottom of the ocean, the flat-fish of long ago needed to keep an eye on what was happening above, and so it rolled on to its side. One eye was keeping watch upwards while the other eye was staring, uselessly, at the ocean floor. Evolution favoured both eyes pointing upwards, and that is

Fig. 3.4 Asymmetric creatures: the antelope with its helical horns. Roland Seitre/Still Pictures.

what the modern flat-fish is like. The result is an asymmetrical fish: flat, with two eyes on one side and none on the other.

It turns out that although individual species of flat-fish are asymmetric, left rollers or right rollers, overall both left and right versions occur. For example, there are several varieties of flounder. In the American Atlantic there is the "summer flounder" whose eyes are usually on the left side, whereas the "winter flounder" has them on the right. For an individual fish the asymmetry is an advantage but nature has no preference for the right or left versions. As with the crossbills, nature originally broke the symmetry but it did not matter which way.

Lobsters have one little limb and one big one; the small one is for clutching and the large for tearing. It is advantageous to have these different limbs for these two tasks and so nature has created the asymmetric lobsters, but with no preference for left and right. There are lobsters where the little limb is at the left, while for others it is to the right. Here again, the job to be done imposes the asymmetry on the creatures individually, but whether it chooses left or right is accidental.

Why do mirrors reverse left and right but not up and down?

The short answer is: they don't!

At the start of this chapter we met Chauncey Gardner watching his image on the television screen going the "wrong way" relative to what he was used to seeing in his mirror. We see ourselves in mirrors, whereas others see us in the flesh. As we comb our hair, guided by our enantiomorph image, we perform the task as a reflex action, learned by relentless repetition over the years. Some time or other you have probably been in a room where there was a mirror on more than one wall and you could position yourself so as to see an image of the one mirror in the other. So used are we to reacting to reflections that it can be hard to manoeuvre into the right position for our image to be visible in such a case. As we try to move into what we think is the correct place our image appears to move the wrong way and keeps slipping out of sight. Eventually you find the spot and can then see an image of your image: this is you as others see you.

Now try to comb your hair, or perform some other task, while watching your image. As was the case for Chauncey Gardner, suddenly everything goes the wrong way—or at least, it does in the "left–right" axis, but the concept of up and down causes no problems.

This gives us a clue that the apparent reversals are illusory. To appreciate this better, try doing the following with a single mirror. Stand with your back to the mirror and arc your head backwards until you can see your reflection. Your head will be inverted and your impression may now be that the mirror has reversed up and down. The actual impression varies from one individual to another and further gives support to the suspicion that mirror reversals are somehow illusory.

You can build on this at parties by first inviting people to agree that mirrors invert left and right, and then have them lie on their sides and ask what has been reversed. They will still say their left and right, but this axis will be the up–down one to all others in the room (assuming that they are still upright). These observers will insist that this axis has not been reflected. We now have two independent sets of witnesses, the verticals and the horizontals, making totally contradictory claims.

We can resolve the paradox and expose the illusion by holding a fancy dress party. The condition for admission will be that everyone comes as a harlequin where the right half of their costume is red and the left, lilac. Everyone in the room will be an RR (red right), LL (left lilac). Now look in the mirror. What you will see is a gatecrasher, someone who does not exist at the party: an LR, RL (lilac right, red left) harlequin. The mirror no longer reflects left and right!

If you are still unhappy, imagine the plane of the mirror being in the east to west direction and place a sign with an arrow and "W" to indicate the direction west. Now turn side on and march off westwards. In the mirror you will see your enantiomorph and a sign also with "W" (courtesy of the precise symmetry of that letter!); as you head off west so does your image. There has been no reversal in the east–west direction and none in the up–down. If we were harlequins we would not suffer from this illusion that mirrors reverse left and right.

So what do mirrors do? The answer is that they interchange front and back: the axis perpendicular to the mirror reverses. Starting from the wall on which the mirror hangs, the room comes out towards you and continues back beyond you. Viewed in the mirror, the room heads away from the real you towards your image and continues "forwards" beyond it. The illusory "paradoxes" come about because you incorrectly imagine yourself having turned around, as if to become your mirror image. However, were we harlequins we would immediately realize that we could not

Fig. 3.5 Mirror asymmetric harlequin.

reach the creature behind the mirror; the mirror world is essentially different from ours.

Art and artefacts

Appreciation of symmetry is profound and pervasive; it permeates art and architecture. To the mediaeval mind, symmetry and God were synonymous.

I once saw a beautiful Byzantine silver tray, depicting the Last Supper of Christ and the Apostles, which preserved the symmetry ingeniously. To left and right of Christ there were identical numbers of symmetrically structured Apostles. The artist of the time preferred to draw side views; the problem was how to draw Christ side-on while maintaining the perfect symmetry that befitted such a holy relic. In a bizarre construction the artist invoked two Christs at the Supper—one facing the left and the other, the right.

While symmetry was *de rigueur* in Byzantine relics, other cultures have taken an opposite stance. Japanese temples deliberately include asymmetry so as not to encroach on the perfection of the gods. The idea that symmetry is suitable only for the Lord but not for human artefacts has led to asymmetries being built deliberately into some christian cathedrals. If you visit Notre Dame in Paris, at first glance it may appear to be mirror symmetric, but on closer inspection, you may be able to tell that the towers are 2 metres different in height and the portals are different in design too.

Once one is alerted to asymmetries they show up everywhere, even in places where one has been looking for years, but remained blind to them. As a young boy I sang in the choir at Peterborough Cathedral. Almost every day for a period of six years I would walk across the Minister Precincts from rehearsals in the choir school

Fig. 3.6 Byzantine image of "The Last Supper," with two images of Christ preserving the perfect symmetry.

to evening service, marvelling at the west front of the Cathedral. What has been described as one of the finest frontages in Europe is, in essence, three huge arches, each topped by a gable, while at the southern and northern ends of the frontage are identical spires. Superficially the whole is a standard example of symmetry in Gothic architecture. Step back from the portal and you will see the bases of two towers behind the canopy, north and south of the east to west axis of the nave, symmetrically placed. However, this is where the symmetry ends as only one of the towers has been completed. This asymmetry is part of the aesthetic appeal of the whole, giving the west front a special character that exact symmetry lacks.

Once asymmetry shows up, one's immediate reaction is to wonder what disturbed the "natural" order. The bases of the twin towers are symmetrical which suggests that something untoward happened during the building that interrupted the completion of the symmetrical pair. The story that has been handed down from one generation of choristers to the next is that the cathedral is in danger of falling down. According to this tale, it was after building the northern tower that the mediaeval masons discovered that the west front was having difficulty taking the strain. It leans out from the vertical by nearly half a metre from top to bottom and a second tower could have caused the western end of the nave to collapse. Building the second tower, by the rule-of-thumb engineering of those days, was thought to be too risky. Another story is that the money ran out.

At Chârtres the symmetry is broken more dramatically; the west front has the perfect rose window, as in Notre Dame of Paris, and the north tower is highly ornate while the southern counterpart is jarringly different. It appears that after building the northern tower, the money ran out as in Peterborough. However, at Chârtres the building was completed later. Years had passed before finances enabled the southern tower to be built, but by then fashion had changed—or perhaps there was only enough cash for a basic model rather than the "super-de-luxe" version that was the northern tower. In any event the symmetry was broken. The reasons for the asymmetry are understood but it was chance that led to the left or right tower being affected.

Another famous asymmetry is the twisted spire of Chesterfield church; this is a result of too much seasoned timber having been used. The pillars on Chesterfield station have a left-handed spiral; a mural in the old waiting-room on platform 2, however, showed the pillars as right-handed.

Fig. 3.7 Asymmetry of Christian cathedrals: Peterborough Cathedral. Martine Hamilton Knight/Arcaid.

Asymmetry has on occasion been incorporated for much baser motives as when French counterfeiters exploited it in order to escape the guillotine. In the 1980s the French 10-franc coin had a fleur de lys on one side and a face on the other; engraved on the side were "Liberte, Egalite, Fraternite." To read these words it was necessary to hold the coin horizontally with the face uppermost. This coin has now been replaced following the counterfeiting of millions of them. Producing exact copies of the coinage was, by an anachronism, still a capital offence and so to avoid this the forgers made a subtle change. They produced the forged versions with the "Liberte, Egalite, Fraternite" inverted such that to read the words the coin had to be held with the face at the base. These coins were immediately recognizable once one was alerted to their characteristic chirality and, in the south of France, were so common that they were accepted as a valid currency.

The use of the word "clockwise" itself reminds us of an established asymmetry. Clocks are usually right-handed (in the sense that the hands move "clockwise"); left-handed clocks would work just as well. In Prague there is a synagogue with both kinds of clock; the left-handed one (whose hands move anticlockwise) has Hebrew lettering on it. This reflects the other mirror asymmetry that is so common that I haven't mentioned it—every line in this book is written from left to right.

If you want to enjoy more of the paradoxes and puzzles of the mirror world, read *The Ambidextrous Universe* by Martin Gardner (no relation to Chauncey Gardner) who has devoted an entire book to this topic.

Chapter 4

Mirror molecules and the origins of life

I was sitting outside a restaurant in Spain one summer evening, awaiting dinner. The sun was setting and daylight slipped away below the horizon, the air became still, and the aroma of the kitchens tantalized my taste buds. My future meal was coming to me in the form of molecules drifting through the air, too small for my eyes to see but detected by my nostrils. The ancient Greeks first came upon the idea of atoms this way; the smell of baking bread suggested to them that small particles of bread existed beyond vision. The cycle of weather reinforced this: a puddle of water on the ground gradually dries out, disappears, and then falls later as rain. They reasoned that there must be particles of water that evaporate, coalesce in clouds, and fall to earth, so that the water is conserved even though the little particles are too small to see. My paella in Spain had inspired me, four thousand years too late, to take the credit for atomic theory.

The ancient Greeks decided that the world around them was made of four elements: air, earth, fire, and water. The more modern atomic theory came about with John Dalton, a Manchester philosopher, early in the nineteenth century. He argued that everything was made of very tiny objects which are joined together to build up the substances that are large enough to see. The modern elements are the atomic ones, including hydrogen and oxygen that are gases at room temperature, liquid mercury, and solid gold. The various elements have different values (do you prefer gold or carbon?) and can occur in different forms (carbon may be graphite or compressed into diamond—so choose your answer to the previous question carefully). Silicon is uniquely suitable for modern electronics. The malleability and toughness of lead, the conductivity of copper, and the lustre of gold are but some of the specific attributes that give each a special character. In combinations they form what are known as molecules; from some hundred atomic elements there are countless molecular combinations that form the substances of the universe.

The ways that these atomic elements can link together are understood well enough that with modern computers it is possible to design new combinations and create substances in the laboratory that may never have existed before. This is common in drug manufacture. It has been calculated that in principle there are more than 10^{100} different combinations of drugs that could be made—a huge number. (In fact it is more than a billion times greater than the total number of atomic particles that exist in the entire universe—so drug companies will have to be selective!) What is to me more remarkable is that all of life, be it human, animal, insect, plant, is based principally on just four elements: carbon, hydrogen, nitrogen, and oxygen. For our nervous system to operate it is important to have traces of other elements and compounds, such as salts to help carry electrical current around our organic circuitry, but the essential richness is governed by these four. And of these four it is carbon that is the key element. Carbon forms so many possible linkages (and essentially all of those 10^{100} potential drugs involve carbon in them) that chemistry has been split into two broad fields: "organic" chemistry (which deals with carbon compounds) and "inorganic" chemistry (which covers the rest).

If we mix hydrogen and oxygen gases together and then disturb them with an electric spark, they will combine with great enthusiasm to form water. Two H(ydrogen) atoms and one O(xygen) atom form a single molecule of water—hence the famous chemical description of H_2O. There are many ways to find out the constituents of other compounds. Sugar is usually eaten but if instead we burn some in the presence of oxygen, the sugar will break down into water (H_2O) and carbon dioxide (CO_2). If we weigh the amounts of each—the sugar, the oxygen, the carbon dioxide, and water—we would find that the sugar initially contained 12 carbon atoms for every 11 combinations of H_2O. The French chemist Antoine Lavoisier, in 1784, realized that substances such as these always combine in fixed proportions (he has become thus immortalized as the "Father of Chemistry") and so the simplest chemical combination for sugar is 12 carbon + 11 lots of H_2O giving a grand total of carbon (C) 12, hydrogen (H) 22, and oxygen (O) 11—hence the chemical formula for sugar: $C_{12}H_{22}O_{11}$. This sugar molecule is part of a general family known as carbohydrates, so named because they contain carbon (the "carbo") and water ("hydrate").

The individual atoms join together like Legos, building up complex chains and structures such as the sugar or even the DNA in our genes. The atoms of any element have a certain number of "hands" that can join one to another (what these hands are need not concern us until Chapter 6); hydrogen atoms each have one hand, oxygen has two, and carbon has four, for example. It is the number of hands that determines the shape and symmetries of the structures and it is the four-handedness of carbon that enables such a variety of possible conjunctions either of carbon atoms on their own or with atoms of other elements. Chemists draw cartoons to illustrate the linkages, with solid lines representing the joined up hands of two neighbours. The "fourness" of carbon and the "oneness" of hydrogen immediately allows a rich sequence:

Methane, ethane, propane, butane, pentane, and hexane are some of the names denoting the increasing numbers of carbon atoms in the molecules, and they are so common that many are household words already. You can play games seeing what forms you can make joining carbon with various elements while obeying the 1:2:4 rules for H, O, and C and allowing yourself to use all three dimensions.

The simplest example of the series is a single carbon atom surrounded by four hydrogen atoms. The chemical formula is CH_4 and it represents a single molecule of methane or marsh gas. The form in Fig. 4.1 is drawn in a plane but if you think about the symmetry in three dimensions a tetrahedron emerges consisting of a carbon atom in the centre, equidistant from each of the hydrogen atoms at its corners; the distance between any pair of hydrogen atoms is the same. This configuration is as symmetrical as can be in the circumstances; rotate it in either direction through 120 degrees around the axis connecting any hydrogen to the central carbon and the molecule looks the same. It also appears identical to its mirror image.

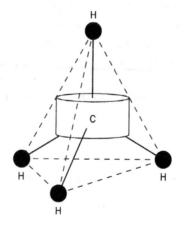

Fig. 4.1 Three-dimensional structure of the CH_4 molecule.

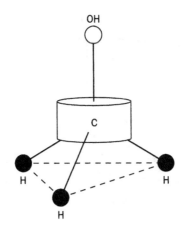

Fig. 4.2 Three-dimensional structure of the CH$_3$OH molecule. Rotate about the vertical by 120 or 240 degrees and the picture remains unchanged.

To visualize the structure, think of the central carbon atom as a black snooker ball surrounded by four red balls representing the hydrogens. Now replace one of the reds by a green ball. If you suspend this model by the green ball it is symmetrical when rotated through 120 or 240 degrees, but suspended by any other ball the rotational symmetry is lost (as in Fig. 4.2). If the green ball is in the plane of the mirror there are three orientations of the mirror where the model is identical to its mirror image. In the real world think of the red balls as hydrogen and the green as an oxygen and hydrogen joined together. (This combination is known as a hydroxyl radical.) The oxygen, (O) has two hands, one linking to the central carbon and the other to hydrogen. (Its shorthand label is "OH"). The resulting molecule is CH$_3$OH, (Fig. 4.2), and is methanol or wood alcohol.

Now change another red ball, this time replacing it with a yellow. This gives a black at the centre surrounded with two reds, a yellow, and a green. Chemically the yellow could be the combination CH$_3$ with the others as before. The molecule is C$_2$H$_5$OH

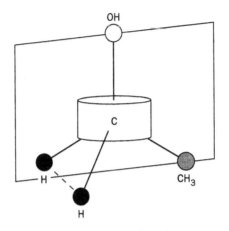

Fig. 4.3 With two hydrogen atoms on the legs, the C_2H_5OH molecule is not symmetric under rotation but remains the same when reflected in a mirror bisecting the two hydrogen atoms.

(known as ethyl alcohol or grain alcohol). The tetrahedron (Fig. 4.3) has lost all rotational symmetry but can still be superimposed on its mirror image (the mirror plane contains the central carbon and bisects the two red balls, or equivalently the two hydrogen atoms).

But now suppose that all four corners have different colours—green, yellow, red, and blue (instead of the second red). There are no planes of symmetry. Suppose that we suspend the model by the green ball as before and also build it such that the red and blue are respectively on the right and left side of the base tripod. Now view the model in a mirror which is parallel to the red and blue balls. You will see the image having the red ball on the left and the blue ball on the right. This change is analogous to the example we met in Chapter 3 of the party guests wearing harlequin costumes with red on the right and lilac on the left. The harlequin in the mirror was distinct and recognizable as not being a party member. In similar vein, the tetrahedron in the mirror is distinct from that in the original reality (Fig. 4.4).

What might the coloured balls represent in the world of organic molecules? We have already met carbon (C), hydrogen (H), and the template combinations CH_3 and the hydroxyl (OH). There is also a chain COOH (the "carboxyl" combination which is acidic) and NH_2 (which is one hydrogen atom short of becoming an ammonia molecule, NH_3). There is enough variety here to be going on with in making a tetrahedral model. With C in the centre all that we need is four different combinations for the corners of the tetrahedron. Let's choose H and CH_3 as before, and for the others, NH_2 ("amino-") and COOH ("-acid"). The resulting molecule is the amino acid known as alanine. It can occur in two mirror forms which for convenience we call right and left (I will describe later what these are in an absolute sense). The tradition is to use latin prefixes to distinguish them: L for "laevo" (left) and D for "dexter" (right)—hence L- or D-amino acids. Any organic compound incorporating both amino and acidic pieces is called an amino acid.

Complex combinations of amino acids create proteins whose general structures can be compared to the tetrahedral example. The carbon is partnered by a H(ydrogen), by the amino NH_2, by the COOH, and, in place of the CH_3 there will be some other group of C, H, N, O varying in composition and structure which is called a "side-chain." Long chains of amino acids may join together in this side chain. Many common proteins contain over a hundred amino acids.

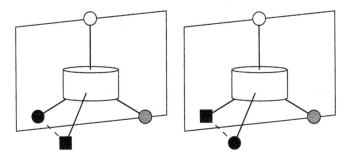

Fig. 4.4 This simple amino acid has no mirror symmetry and can exist as two forms that are mirror images of one another.

About one-sixth of our body weight consists of proteins which form the bulk of our muscles, skin, and other tissues. The amino acids are the templates of proteins and are essential to life. In the entire animal kingdom there are less than 25 different amino acids of which we can make 10 ourselves, but the others we have to obtain from our diet, the richest source being the meat of other animals. The essential amino acids can occur in either L- or D-forms, however life on earth almost exclusively uses the left-handed form. How and why is one of the questions we need to pursue. First we need to specify what we mean by L- and D-types as the occurrence of left and right mirror molecules is much more widespread than just amino acids and underpins the nature of life.

Polarized light

Louis Pasteur's name will always be associated with pasteurization. *Bon viveurs* will also thank him for saving the French wine industry; while many of us will have benefited from his development of vaccines. All of these were the result of his work with micro-organisms in the latter half of the nineteenth century. He found that passing certain micro-organisms through animals or exposing them to air or varying the culture, weakened them: this was a first step to producing vaccines. Around this time also, the French wine industry was being undermined by hostile organisms and Pasteur's work resolved that problem too. He showed that it was micro-organisms rather than spontaneous changes that caused milk to ferment and produce lactic acid; "pasteurised" milk was the result of this knowledge. These advances were all made when his scientific reputation had already been established; it was his earlier discoveries in 1848, at the age of 26, that have had the most profound consequences for structural chemistry, the nature of life, and modern drug manufacture.

When grapes ferment, tartaric acid forms, as in the crystalline deposits found at the bottom of wine bottles. Industrial processes in the Alsace region had been found to produce a chemical that was identical to tartaric acid, in that it contained four carbon atoms, two hydroxyl, and two COOH groups, but with other mysterious differences. Whereas the usual tartaric acid melted at 173°C, this new form melted at 210°C. The novel form was named racemic acid ("racemus" is Latin for a bunch of grapes) and the puzzle was how two apparently identical chemical compounds could be physically so different.

The two forms of the acid also affected light in different ways. As this turned out to have profound implications for our understanding of the nature of life, let's take a brief detour into the subject of polarized light and crystals.

Light waves consist of electric and magnetic fields vibrating back and forth transverse to the direction of the beam. (There's more on this in Chapter 5.) This vibration normally takes place in all planes that pass through the line of the beam's motion. The regular arrangements of molecules in some crystals make the substance transparent only to light that is vibrating in one plane, while being opaque to all others. The light emerging from the far side of the crystal is therefore vibrating only in one plane: it is said to be "plane polarized." Polaroid lenses pass only one plane as you can verify by holding two pieces one behind the other and gradually rotating one through a right angle relative to the other. As you do so you will see the view go dark as the preferred planes of the two polarizers become misaligned.

When plane-polarized light passes through quartz, more interesting things happen. Quartz crystals have an asymmetric shape that is distinct from its mirror image. The crystal may be in either of these forms. Depending on which of the two this is, the plane of the polarization of light passing through the crystal rotates in either a clockwise or anticlockwise direction. These effects come about because of the asymmetric arrangement of the individual molecules within the overall lattice of the crystal, and not from any asymmetry of the individual molecules. This is in contrast to organic compounds where the mirror asymmetry is intrinsic to their basic molecules.

So having made that detour, now let's return to the mystery of the two forms of tartaric acid. Not only did they melt at different temperatures but Jean Baptiste Biot, the leading French physicist, had discovered that the tartaric acid rotates the plane of polarized light to the right, whereas the other form, "racemic acid," does not. Biot suggested that the cause lay in the arrangement of the individual atoms and that the rotation of the light was because the molecule was mirror asymmetric in the same way as the crystals had been. This was astonishing as it would not be for another 30 years, in 1874, that the idea of the tetrahedral carbon atom and its ability to create mirror asymmetric structures was proposed by Joseph Le Bel in France and, independently, by Jacobus van't Hoff in Holland. Pasteur took up Biot's idea and proved it correct by some meticulous experiments.

Pasteur began with racemic acid—evaporated it and let it crystallize. With tweezers he carefully separated the crystals into piles of right-handed and left-handed forms and then he dissolved each pile in water, creating two separate beakers of acid, one for each handedness. Pasteur was now ready to make the critical test of seeing what happened when polarized light passed through each liquid. When the light passed through the liquid derived from the right-handed crystals, it behaved the same as when it encountered conventional tartaric acid: it rotated to the right. Then he repeated the experiment with the other beaker. Here he made his great discovery; this liquid rotated the light by exactly the same amount but in the opposite direction. Here was something that had never been seen before, a form of tartaric acid that rotated light to the left. This suggested that the original racemic acid (which did not rotate light at all), was in fact an equal mix of the two mirror forms, whereas the familiar light-rotating tartaric acid was only the right-handed form.

Pasteur was fortunate that he did these experiments in the winter. At temperatures above 28°C the crystals of racemic acid change from right- and left-handed forms into mirror symmetric structures that would have negated the whole effect. Figure

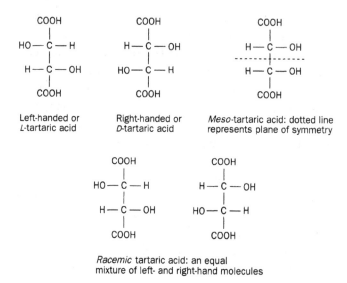

Left-handed or
L-tartaric acid

Right-handed or
D-tartaric acid

Meso-tartaric acid: dotted line
represents plane of symmetry

Racemic tartaric acid: an equal
mixture of left- and right-hand molecules

Fig. 4.5 Left- and right-handed tartaric acid. There are two ways of making symmetrical forms. One is to have a 50:50 mixture of both—racemic acid (racemic is used today in general to denote such equal mixes of two mirror forms). However, one can twist the upper or lower half to create "meso"-tartaric acid molecules which are mirror symmetric about the plane shown by the dotted line.

4.5 shows the molecular structures of tartaric acid, and illustrates both the left- and right-handed forms and the subtly different ways they are realized.

Moulds, such as those that give rise to penicillin, eat tartaric acid only when it is the right-handed variety. Pasteur discovered that when such a mould was put in a tartaric acid mixture, the mould would grow until all the right-handed acid was used up: the remaining acid would be pure left-handed. This was possibly the first example of a biological machine, where a living mould was being used to create left-handed acid. It also gave the first hint that the mirror asymmetry in molecules controls our lives.

Lactic acid in milk is mirror asymmetric and its molecular structure is shown in Fig. 4.6. (There are two lines joining the

C(arbon) and O(xygen) because oxygen has two links, satisfying the four-link-rule for the carbon neighbour). The molecule can occur in mirror-reflected L- and D-forms. Milk contains one of these whereas the other is produced in our muscles by the kind of hard exercise that makes them ache. When contemplating life in the looking glass world, Alice wondered if mirror milk would be fit to drink. By a mirror person it would, but probably not by us; I certainly am not attracted by the thought of drinking "muscle-ache" lactic acid.

Handedness pervades the living world and has some bizarre manifestations. We can smell it and taste it, while forensic science and whole pharmaceutical industries rely on it.

The dentine in teeth contains aspartic acid whose molecules can occur in mirror-image forms. In childhood, only the left-handed form is present, but over the years it gradually changes to its right-handed mirror image, the relative amounts of the two showing the age of the person. Forensic medicine can use this to determine the age of a human corpse.

Our noses are made from proteins that are mirror asymmetric and can themselves smell the difference between left- and right-handed versions of the same molecule. As my experience in smelling the cooking of paella reminds us, the nose can detect things smaller than our eyes can see; the way things smell depends on the shape of the molecules and the way they fit into the receptor molecules in the nasal olfactory hairs. Smelling molecules of carvone in its right-handed form gives us the recognizable odour of spearmint, whereas the left-handed form gives the smell of the caraway

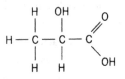

Fig. 4.6 Lactic acid molecule.

plant. The left- and right-handed forms of limone distinguish the smell of oranges and lemons.

We can taste mirror molecules. The glucose molecule in Fig. 4.7 is asymmetric and rotates plane-polarized light to the right. Two of these D-glucose molecules link together to form sugar; it is often called dextrose because of its right-handed nature. Fructose, by contrast, has the same atoms but in the left-handed form, which is the reason for its alternate name of levulose.

Our digestive system responds in different ways to left- and right-handed versions of the same molecule. One may be digested, turning into energy or tissue, while the other may be excreted without otherwise being processed. We are in effect machines for processing handed molecules. This is what Pasteur had discovered with his mould and organically we are but more complicated versions of mould. Not only do we process left and right differently but they can trigger different responses in us. For example, dextrose is less harmful to diabetics than is its mirror image, levulose. This differing response to mirror molecules requires great care when designing drugs—a detail that is now well recognized but had tragic consequences in the case of the drug thalidomide.

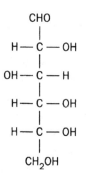

Fig. 4.7 D-glucose, which links with another D-glucose molecule to form sugar.

Thalidomide had been used in medicine as a sedative without any problems until it was given to pregnant women. Between 1959 and 1962, in Germany and Great Britain, around 3,000 babies, whose mothers had been taking thalidomide during pregnancy, were born deformed. Subsequently it was realized that the clinical trials had not adequately appreciated the different effects that right- and left-handed versions of the substance could have. It was withdrawn from use. Today, with chirality in drug manufacture much better understood, thalidomide is having a resurgence as a means of treating forms of leprosy.

Carbon and life

We met carbohydrates earlier in this chapter. As their name records, these are molecules built from carbon and water. They are the fibre in our diet and form the cell walls of plants. The simplest example involving one carbon atom is CH_2O—formaldehyde. Figure 4.8 shows how it forms and is mirror symmetric. More and more carbon atoms and associated water molecules create a rich family of carbohydrates. A simple template is $H—\overset{|}{\underset{|}{C}}—O—H—$ where the unused hands on the carbon can join to neighbouring carbon atoms and hence in turn to a carbohydrate template. These larger structures are mirror asymmetric in general.

The OH (hydroxyl) group sticks to other atoms and molecules very strongly. The wetness of water is an example of this stickiness, where the OH has linked to a single H(ydrogen). Glucose is one of many examples of six C-HOH templates; it is covered in OH groups which enables it to stick onto other molecules and transport energy around our bodies (we have about 5 gm in our blood at any one time). Its stickiness is familiar in another context. If we heat glucose with a little bit of water, the glucose molecules adhere together, forming a huge molecule—toffee. The stringing of glucose in long chains forms many giant molecules with familiar

Fig. 4.8 Formaldehyde molecule.

names like starch, cellulose, and glycogen. Cellulose is long and straight; plants worldwide make a billion tonnes of it each year simply to keep themselves upright. Our digestive system cannot process it and it forms the roughage in our diet.

It is this ability of carbon to create long chains that is the essence of the complexity that we term life. I started this section with amino acids and proteins and it is to these I return now, as they bring us as near as science has yet come to explaining the mystery of being.

Every life-form on earth contains proteins in some shape or form; every cell in our bodies contains hundreds of different proteins, there being over 100,000 varieties in all. If these are likened to the words in a dictionary, then the letters from which they are constructed are the amino acids. As 26 letters make the English alphabet, so a similar number of amino acids make our proteins. The 26 letters can each be built up from vertical, horizontal, and diagonal bars as in alpha-numeric displays; the amino acids also are built from pieces, each amino acid having an amino end and an acid end with assorted C(arbon), H(ydrogen), and O(xygen) atoms in between. The "amino-end" and the "acid-end" of two amino acids attract one another as do the north and south poles of two magnets. The act of sealing involves the loss of water: a H(ydrogen) is taken from the amino end and a hydroxyl combination (OH) from the COOH to give the H_2O. What remains at the amino end is NH, while at the acid end only CO survives; it is these NH and CO combinations that stick the neighbouring amino acids to one another by what is known as a "peptide" link.

Each amino acid has a left or right asymmetric form. As more and more of them join one another, the resulting chain will twist in a direction determined by the asymmetry of its constituent amino acids. To visualize this, think of a spiral staircase. Each step is asymmetric, with a fat outside and thinner inside. It is this common asymmetry in the steps that makes the spiral when they are joined one to the next; a series of thin right-hand ends makes a right-hand spiral. A similar geometry acts when asymmetric amino acids link together. Each left-hand amino acid joins to another. Look along the backbone, up the staircase. The resulting spiral moves left (anticlockwise) as it comes towards you, or right as it goes away. With our conventions from screws, we therefore say that the resulting protein is a right-handed coil.

The number of combinations of amino acids possible in forming proteins is hard to comprehend. The 26 letters of the alphabet produce all the words in the dictionary. Including those from other languages, and the additional "allowable" sounds that could qualify as words (allowable in the sense that vowels are needed on average every three or four letters to make an utterable sound), we find about 10 million pronounceable "words" with up to six component letters. There are legitimate words with more letters than these; in German, the conjunction of words to build up macrowords is akin to the organic creation of huge molecules. There are billions of varieties of proteins that can exist, at least in theory, each with its own unique properties. It is this huge variety that provides the wherewithal for the special tasks that complex life-systems require; carbon's great affinity for teamwork enables this. Evolution provides the means of "accidentally" forming them and thousands of millions of years provided the time for evolution to enter the lottery—sufficient time in which winning solutions could periodically be chanced upon. Had Darwin not come up with the idea of evolution from his observations of the species, modern biochemistry and microbiology would have deduced it as knowledge on the structure of proteins and, more fundamentally, DNA emerged.

DNA

The essence of life is carbon's ability to form complex structures. The moment that a variety of pieces join to the carbons, asymmetric structures emerge. Left- and right-handed versions occur and when linked to one another, spirals form. The asymmetry of the steps (whether thicker at the left or right end) determines the direction of the spiral staircase; the structure and asymmetry is more complex in the case of amino acids but the principle is the same. The shapes determine the spiral.

The shapes of the steps are those of four different amino acids named adenine (A), thymine (T), guanine (G), and cytosine (C). Before Crick and Watson solved the structure of DNA, chemical analyses had shown that the amounts of A and T were the same, as were G and C, while X-ray images of DNA crystals had shown that the molecules had a helical structure. The problem was how it all fitted together.

As James Watson describes in *The Double Helix*, the puzzle began to be solved the moment that he realized that when the A and T amino acids join together, their overall shape is identical to that formed by G and C. It is this perfect symmetry in shape that is the key. If adenine always pairs with thymine while guanine always pairs with cytosine, a series of identically shaped "steps" emerges. It was only after Crick and Watson had built precise models of the shapes of the CGAT pieces forming the steps that they saw that the resulting coil is right-handed.

In fact DNA is a double helix, as the title of Watson's book records. In structure it is like a ladder that is twisted with the CGAT molecules pairing up to form the rungs, while the two helices are the sides of the ladder. Replication occurs when the middle of each rung is cut, exposing a single CGA or T; the unique pairing then imprints the matching of these onto a new set of partial rungs in sequence GCT or A.

It is the asymmetric shapes of the molecules forming the rungs that forces the DNA to be twisted; our DNA is right-handed. In a mirror, all the component parts would be reversed and the mirror DNA be left-handed. Humans do not use this; right-handed DNA for all is the rule. Evolution plays an essential role in this. As only like spirals can link to make a double helix, there is no advantage if some of us had right and others left, or if we had a mix of both. It appears essential for effective procreation, that the half provided by the male matches that from the female, and the most efficient way is if they have only one and the same handedness. While this does not explain how and when the asymmetry in the amino acids of our DNA originated, evolutionary advantage aided by the vastness of time could be the cause.

Alternatively, left-handed amino acids might have come here from outer space, so that we are all descended from aliens. This idea, beloved by science fiction, might even be science fact. In 1969 a meteorite fell from the heavens and landed near Murchison in Australia. It turned out that the "Murchison meteorite" contained an extraordinary number of organic molecules and, moreover, the amino acids within them were left-handed.

Could meteorites and comets have carried the characteristically twisted molecules to earth, which served as the building blocks for proteins? Once here, these asymmetric molecules could have helped form the right prebiotic soup from which more complex organic molecules arose. The fact that left-handedness in amino acids is not unique to life on earth is tantalizing to say the least.

While the discovery of left-handed amino acids in the Murchison meteorite suggested that the mirror asymmetry may have arisen in outer space even before our earth had been born, it did not answer how it had come about. However, astronomers have found a new clue that may explain it.

Pasteur had found long ago that mirror asymmetric molecules will rotate polarized light one way or the other, depending on their left- or right-handed form. Laboratory experiments in this century have shown that circularly polarized light, where the electric field is rotating, can selectively destroy either right-handed or left-handed molecules oriented in its direction of rotation. Recently astronomers have discovered that there are surprisingly large amounts of circularly polarized light streaming anticlockwise from regions of the Orion nebula where new stars are being born and organic molecules are also known to be present. The conditions are believed to be similar to those that were prevalent when our own solar system was being formed.

So it is plausible that left-handed organic molecules were the survivors of bombardment by intense polarized starlight at the dawn of the solar system. As a preferred handedness in molecules is believed to have been necessary for the origin of sustained life to have been possible, the astronomers' findings suggest that our planet's suitability for life may be as much due to the environment in which our solar system formed as to the local conditions in the early earth. If this is the case, then organic molecules, precursors of life, that may be present on other bodies in the solar system, should also have a left-handed form.

The question of how conditions on earth stirred life in inanimate organic molecules remains unanswered, but these discoveries bring us one step closer to the answer. That we have been able to unearth these clues is due to the forensic application of scientific discoveries on the nature and make-up of matter that were made almost exactly a century ago—of X-rays, radioactivity, and atomic structure. X-rays enable us to see directly the structure of crystals such as DNA; radioactivity allows us to trace the origin of amino acids and, with nuclear physics, provides us with a means of distinguishing chemically identical forms that originate on the earth from those of outer space. In the next three chapters we will see how this cosmic detective work was born.

Chapter 5

Unearthly visions

The apparition was so awful that Wilhelm Roentgen wondered if he had taken leave of his senses. He could hardly have been more surprised if he had looked into a mirror and no reflection had stared back. He let go of the metal that he had been holding and jumped, startled by the noise, as it hit the floor.

It was approaching midnight on 8 November 1895. For some time scientists had been reporting bizarre apparitions when they electrified the thin gas in vacuum tubes. The English physicist William Crookes, who saw unearthly luminous clouds floating in the air, had become convinced that he was producing ectoplasm much beloved of Victorian seances and had turned to spiritualism as a result. In Germany, Roentgen was doing similar experiments and now, alone in the night, his imagination ran wild.

Earlier that day, as the November dusk darkened the laboratory, he had noticed that whenever he made sparks in the tube, a fluorescent screen at the other end of the laboratory table glowed slightly. This was the signal that he had been looking for, the sign that invisible rays were being produced in the spark tube, crossing the room, and striking the screen, producing the faint glimmer. If that had been all it would have been enough, but after a late meal Roentgen had returned to the laboratory where in the darkness the glow was easier to see and it was then that the phantom happened.

To track the rays he had been putting pieces of card in their way, but the screen continued to glow whether the cards were there or not, as if the rays were able to pass clean through them. He then tried to block the rays with metal but thin pieces of copper and aluminum were as transparent as the card had been. Somehow Roentgen's electrical device was producing rays that seemed to be impervious to matter, but at last he found something to stop them: lead left a shadow, proving that the mystery rays were definitely real.

He moved the piece of lead near to the screen, watching its shadow sharpen, and it was then that he dropped it in surprise: he had seen the black shape of the metal held by the hand of a dead man. Pulling himself together he slowly opened his clenched fist and looked astonished once again at the dark skeletal pattern of the bones as his hand moved across the face of the screen. Still doubting what he saw he took out some photographic film for a permanent record. Roentgen had made one of the most monumental discoveries in the history of science—X-rays—and seen for the first time images that are today common in every hospital.

Six weeks later, on the Sunday before Christmas, he invited his wife Bertha into the laboratory and took a shadow graph of the bones of her hand, with her wedding ring clearly visible. This is one of the most famous images in photographic history and propelled him within two more weeks into an international celebrity. The medical implications were immediately realized and the first images of fractured bones were being made by January 1896—even though none yet knew what the mystery rays were.

If any single moment marks the start of a modern physics and science it is that Friday evening of 8 November in 1895. But why then and not before? Why Roentgen and not any of half a dozen others? As is so often the case, a seminal discovery was missed by others who had been better placed or had even seen the phenomenon and failed to realize the evidence before them. Discovery involves being in the right place at the right time, being prepared, and being brave enough to break away from the existing mind-set. There can also be unique features to help to single out one individual; in Roentgen's case it may have been that he was colour blind.

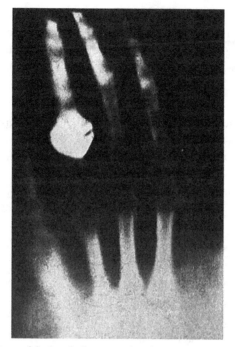

Fig. 5.1 X-ray of the hand of Roentgen's wife.

Let there be light

The steam engine was arguably the epitome of the nineteenth-century human ingenuity, where simple principles of basic physics were turned into a powerful technology. The main ingredients, coal or wood, need no industrial processing.

As fuel they heat water until it boils—the principle being that a small volume of water turns into a huge volume of steam. From here on it is engineering; the steam pressure will push a piston which, via rods, turns the wheels, and hundreds of tonnes of weight can be pulled at 100 kilometres an hour. These were the days of the Industrial Revolution. Mechanics and thermodynamics were the staple diet of physics and the engine of technology.

As smokestacks spewed their chemical ash into the skies and into the lungs, there was no environmental movement to question the price being paid for all this, not least because medical knowledge was still primitive and the electronic mass communication of the twentieth century was still decades away.

If the steam engine epitomizes the old mechanics, it is its progeny, the modern electric motor, that symbolizes the great revolution whose consequences are even today still being worked out. Engines, gears and levers, pistons and pulleys in motion were mechanics applied, but the nature of matter, light, and electricity was a relative mystery. It was Michael Faraday who changed all that.

Faraday was a genius born too soon. The first Nobel Prize was awarded in 1901, by which time he had been dead for over 30 years. Had the prizes been in existence in the nineteenth century, Faraday would probably have won no less than six of them. Among his several seminal insights he proposed that electric and magnetic fields permeate space and that these push and pull electrically charged objects and magnetized items respectively. For example, it is the earth's magnetic field that holds compass needles in line, pointing in the direction of the earth's remote magnetic poles.

A great popularizer of science, his lectures at the Royal Institution contained remarkable demonstrations setting a tradition that has lasted over 150 years. When I was invited to give their annual lectures for young people at Christmas 1993, it was with trepidation that I discovered that Faraday had given them on several occasions up to 164 years earlier. In one of these he had made the magnetic field "real" by sprinkling iron filings near a magnet whereby they clustered into patterns forming a topographical map of the field; today, this is a standard part of the school physics curriculum. His lectures were so popular that the horses and carriages collecting the audience afterwards created chaos in Albermarle Street.

Today, single horsepower carriages have been replaced by multi-horsepower motors which are themselves a progeny of one of Faraday's great discoveries: that electric and magnetic fields interweave and affect one another. If a magnetic field is changed, for example by moving the magnet that is its source, it will create *electric* forces. This phenomenon is called induction and underwrites the technology of electric generators. The idea is that changing magnetic fields gives rise to electric fields and vice versa, which is an integral part of the electric motor.

James Clerk Maxwell encoded the phenomenon mathematically in his famous equations. Here we have one of those uncanny properties of mathematics whereby when linked with a limited set of experimentally established facts, it can predict successfully a whole symphony of new phenomena. His equations summarized that a changing electric or magnetic field was vibrating, in the sense that uphill and downhill, say, were interchanged N times each second, the resulting induced magnetic field would also be oscillating at the same frequency; in turn, this induces a pulsating electric field. The whole conglomerate of oscillating electric and magnetic fields would propagate across space as a wave. From that moment, electric and magnetic fields were no longer regarded as two separate entities but became united: the concept of the electromagnetic field and electromagnetic waves was born.

Faraday's measurements of electric and magnetic phenomena provided the essential data that enabled Maxwell to calculate the speed at which the waves travel. He found it to be 300 million metres each second, independent of frequency. This is also the speed of light, which gave Maxwell the essential clue that light and travelling electromagnetic waves are one and the same. Light, the rainbow of colours, consists of electromagnetic waves whose electric and magnetic fields oscillate hundreds of millions of millions of times each second, the distance between successive crests in intensity being in a narrow range around a millionth of a metre. Maxwell's insight implied that there must be other electromagnetic waves, beyond the rainbow, travelling at the speed of light but vibrating at other frequencies.

Infra-red and ultraviolet light had already been discovered. The infra-red rays are electromagnetic vibrations at a lower frequency than our eyes can see, though our skin can sense them as

Fig. 5.2 The founder of electromagnetism: Michael Faraday (as on a £20 note).

heat; and ultraviolet are higher-frequency versions—"infra" and "ultra" referring to the frequencies relative to those of the visible spectrum. These clues impressed and inspired scientists to seek, or to create "artificially," other examples of electromagnetic waves beyond the rainbow. Today's discovery is tomorrow's technology and the new electromagnetic waves were no sooner found than they were being put to use.

In 1885, Heinrich Hertz at the University of Karlsruhe produced electric sparks and showed that they sent electromagnetic waves across space without the need for material conductors—the so-called "wireless." These primitive "radio" waves are electromagnetic waves similar to light but in a different part of the spectrum, Hertz has given his name to the unit of frequency: one "per sec" is called one "Hertz," thousands are kilohertz, and millions are megahertz; so the rainbow corresponds to millions of megahertz. Radio waves are part of the electromagnetic spectrum but far beyond the optical range and the deep infra-red; their frequencies are typically in the kilohertz to megahertz range.

Hertz's discovery of radio helped to establish Maxwell's theory of electricity, magnetism, and light, and in turn led to speculation that there exist "invisible high-frequency rays"—electromagnetic waves that are far beyond the blue horizon. Roentgen set out to find them, and succeeded, for that is what X-rays are—very high frequency electromagnetic radiation. The irony is that at first he didn't realize what he had done. The greater tragedy was that Philipp Lenard had missed the discovery seven years earlier and never recovered from the psychological shock. He degenerated into an anti-Semite, attacking Einstein and, in the Nazi era, authored a four-volume work on German physics which omitted all mention of Roentgen.

Steam engines, ectoplasm and cathode rays

The misadventures had begun in the quest to understand the nature of electricity. Until Maxwell's time the physical sciences had concentrated on mechanical interpretations of natural phenomena. Steam engines, Newton's laws of motion, and the pulleys and mechanization of the mills were the paradigms of the mechanistic perspective of Nature which led to the early Industrial Revolution. With that philosophy it was natural that when eighteenth- and early nineteenth-century scientists visualized lightwaves it was as a mechanical process involving an intangible "ether" through which light propagates—after all, mechanical things need something to mechanize the motion. But by the late nineteenth century, following Maxwell and Hertz, even diehard "mechanists" had to admit that this couldn't be correct. Maxwell had begun a revolution with his realization that light is the result of oscillating electric and magnetic fields in empty space. This inspired scientists to turn their attention to electrical phenomena, which led to a number of discoveries within little more than a year that would define the course of history.

Electricity flowed along wires as if it were a fluid. Connect the wire at one end to the negative terminal of a powerful battery and the other end to a metal plate contained inside a glass tube filled with gas. By this means, electricity passed through the gas and produced its first surprise—the eerie coloured glows that today are refined in modern strip lights and advertising displays.

Thousands of volts and powerful vacuums are needed in order to have any chance of producing X-rays, and these were far beyond technology until the final years of the nineteenth century. Improved vacuum pumps were one of the spin-offs of the railway revolution: the discovery of X-rays and the arrival of "post-steam" technology depended on this apparently mundane fact. As the first steam train chugged along the Stockton and Darlington railway, none foresaw that steam power would play a seminal role in the random walk to modern genetics. These developments were not made with foresight of the eventual discovery of X-rays, but rather opened up more extreme conditions than had ever before been experienced.

When the pioneers passed electric current through gases at lower and lower pressures, a rich variety of effects occurred. First, William Crookes noticed that the nature of the coloured lights changed as the pressure was reduced. In these circumstances the atoms in the gas emitted light much as the Aurora Borealis or "Northern Lights" are stimulated by electrically charged high-energy particles—cosmic rays—hitting the upper atmosphere. The high atmosphere is rarefied, and vacuum technology produces a similar thin gas in the Crookes tubes. The ghostly shimmer of the aurora in a glass tube is eerie even when you know what it is and to Victorian scientists, working in the dark, in all senses of the phrase, it was unnerving. Crookes had become involved with spiritualism following the death of his brother. Seeking a scientific proof of the soul, he became obsessed by the subtle lights in his tubes. Convinced that during seances he had seen "luminous green clouds" and that the lights in the tube were the same as these phantoms, he announced that he had produced ectoplasm.

Although this led to some ridicule, nonetheless his research did reveal the dramatic way that the light effects changed as the pressure dropped. As he reduced the pressure still further, the lights around the "cathode"—the point of entry of the electric current—disappeared, leaving a dark space with the luminous glow reappearing, displayed along the tube. Finally, he removed effectively all of the gas, or as much as his most powerful vacuum pumps were able, and increased the voltage to a maximum. At this the nature of the discharge changed entirely: the aurora disappeared to be replaced by pencil-thin beams emerging in direct lines from the cathode. These were known as "cathode rays."

This was in the 1880s and it would not be until 1897 that the full mystery of the cathode rays would be solved by J. J. Thomson, whom we shall meet later. To anticipate the answer and to help make more sense of what follows, let me give the punch-line now: cathode rays are streams of electrons which are constituents of atoms and the carriers of electric currents. One of their many properties is in causing phosphors to become luminous, the bright picture on a television screen being a homely example.

When cathode rays in a vacuum are focused onto hard targets they can produce X-rays and if the target is very small it can become extremely hot, as in a "cathode-ray furnace." We now know, from the conditions of Crookes' experiments, that when the cathode rays hit the glass in his tube, although most of the energy from the cathode rays was dissipated as heat, about one part in ten thousand of their energy was converted into X-rays. Too much heat would melt the glass, so Crookes ran the experiments with enough power to produce the gassy light displays but no more than was necessary, for fear of damaging the equipment. The result was that he was actually producing X-rays, but they were extremely faint.

Misadventures

X-rays leave images on photographic emulsions; in Crookes' laboratory, even though the X-rays were very dim, they penetrated boxes of unexposed photographic plates that he had stored nearby and duly energized them. When later he came to use them and found that they were already fogged, instead of associating this with his experiments, he returned the plates to the manufacturer under the mistaken assumption that they were defective.

Crookes was not the only one who might have discovered X-rays years ahead of Roentgen: Philipp Lenard had two chances and missed them both.

In 1888, following Crookes' work, but with Roentgen's success still seven years in the future, Lenard had been looking for electromagnetic waves in the ultraviolet part of the spectrum, beyond the blue horizon of visible light. He was using a cathode ray tube and placed fluorescent crystals just outside the tube's front end. If only he had removed more gas from the tube until he reached the low pressures that Crookes had achieved, he would have been able to use higher voltages to make the cathode rays and this in turn would have led to the emission of powerful X-rays. Such X-rays would have hit the crystals in front of the tube, made them fluoresce and captured Lenard's attention. However, in the conditions in which he actually operated, the experiment produced only ultraviolet light and the rather weak X-rays that lie just beyond the ultraviolet (known as "soft" X-rays, in contrast to the high-energy "hard" X-rays that penetrate easily). These rays were absorbed by the tube and never made it to the crystal detector. So he failed both to find the ultraviolet rays that he was seeking and also the greater prize of X-rays that none had suspected.

By 1893, Lenard was assistant to Hertz and was making cathode rays of much higher energies in gases at extremely low pressures, similar to those that Crookes had used. In such conditions he must have been making hard X-rays that penetrated out from the tube into the surroundings. He did notice that photographic plates were turned to intense black but reported, initially to the Berlin Academy on 12 January 1893, and in a more complete paper that was published early in 1894, that this occurred only when the plates were very close to his apparatus. This is what he had expected to find; he had not anticipated that occasionally photographic plates covered by sheets of cardboard would also become blackened. The cardboard sheets were thick enough to absorb the cathode rays and so it could not be they that were responsible. With the benefit of hindsight we now realize that the culprit was X-rays, but Lenard somehow was unable to recognize the unexpected.

We don't know whether Lenard failed completely to realize that a greater prize awaited or had been disturbed enough by these hints to have intended to follow them up. Any plans that he had were interrupted by the sudden death of Hertz on 1 January 1894. Lenard spent much of 1894 editing and eventually publishing three volumes of Hertz's final scientific works; he had also taken over Hertz's duties as director of the Bonn laboratory and was all but lost to research. Wanting to carry on with his cathode ray work, he accepted a professorship in Breslau only to find that there were no suitable facilities there. He quit and took a post in Aachen where at last he could carry on with cathode rays. By now it was January 1896 and too late: Roentgen's sensational picture of Bertha's skeletal hand was already in the newspapers.

Roentgen was colour blind

While other competitors were missing their chance of glory, Roentgen started late and came straight to the prize. Roentgen had read Lenard's 1894 report and some deep intuition alerted him that there was more than Lenard had realized. So he started studying cathode rays for himself and even wrote to Lenard for advice on his experiments.

As in any activity, practice makes perfect; Roentgen repeated Lenard's experiments, learning the quirks of the apparatus. By the autumn of 1895 Roentgen was confident that he had understood it well enough to set out into uncharted territory in search of the high-frequency rays. His basic idea was to maximize their intensity by using the highest voltages and lowest gas pressures possible, and to look at a fluorescent screen in a darkened room in the hope of detecting any faint glimmer as the rays struck.

Crookes and Lenard had seen the cathode rays induce bright fluorescence in the glass walls of the tubes, so Roentgen blocked out these cathode rays with pieces of card. The high-frequency rays that he was looking for had been predicted to be capable of penetrating paper and card, so he hoped to tease them out by detecting a residual dim glow when they reached the fluorescent screen. Absolute darkness was essential. The critical experiments began on the November afternoon; the greyness of winter helped and the heavy curtains of the laboratory achieved darkness.

First, he had to ensure that there was no stray light. He did so by covering the tube and spark maker completely, then pressed the switch to make the high-voltage sparks and cathode rays, and then checked for any signs of light from the apparatus that would show whether or not it had been completely shielded. The fluorescent screen that he planned later to bring near to the cathode ray tube as a detector was out of the way, more than a metre from the apparatus, while he made this first test. This was his first piece of fortune.

The tube was well covered and no light escaped when he pressed the switch to make the cathode rays. The experimental set-up appeared to be perfect but then he noticed in the corner of his eye a faint glimmer from across the room. He looked over towards it but it was gone like the smile of the Cheshire Cat. He pressed the switch again, and the same thing happened: the tube was completely obscured and perfectly dark but there was a glimmer in his peripheral vision. Then he identified where it was coming from: the distant fluorescent screen.

Go and look at the stars on a clear night. You can see the bright ones easily but some are so faint that you can only see them when you turn away. This apparent oxymoron is actually true and happens because very faint objects only show at the edge of your vision where the eyes' rods rather than cones respond. The rods are not required for sensitive spatial resolution and so several of them are connected to a single neurone. The result is that the peripheral neurones gather light from the rods, accumulating enough to trigger an image in the brain, whereas the cones of the central system remain effectively blind. This is how Roentgen first saw the dim glow, and here too was a further piece of luck: Roentgen was red–green colour blind. Fortunately, this does not affect the rod visual system and furthermore tends to increase the individual's ability to discern shapes and contrasts in compensation for their pigment deficiency. So ironically, Roentgen was especially well equipped visually for the task and aided by the serendipitous placing of the fluorescent screen.

Once his eyes were fully adjusted to the darkness and the glow was more clearly visible, he started the series of tests that ultimately led to the full discovery. He placed sheets of card, a thick book, and even a piece of wooden shelving between the tube and the screen in the hope that invisible rays would cast shadows and so give themselves away, but in all cases the glimmer was undimmed, faint but definite, and correlated with the cathode ray tube being "on." All this had taken place with the screen remote from the apparatus and so he now moved it nearer, finally to its originally intended position a few centimetres from the tube. As he did so, the glow increased, eventually becoming a bright green fluctuating cloud (fluctuating, we now realize, because the X-ray intensity varied a lot due to the primitive electrical contacts then available). The brightness of the screen when close to the tube convinced him beyond doubt that invisible rays were hitting it, even though he had found nothing to stop them and make a shadow. He decided to estimate how powerful they were by moving the screen further and further away to see how far the rays could be sensed. He found that the glow remained visible, though ever fainter, up to two metres away; beyond this his tired eyes could not be sure.

He broke off for dinner. Immediately after eating he was back at work, refreshed. In place of the pieces of card and the wooden board as attempted shields, now he tested sheets of metal to see if they could interrupt the mystery rays emanating from the cathode ray tube *en route* to the fluorescent screen. The rays passed through aluminum and copper as easily as they had through the card and wood previously. Then at last he found something to stop them. With the screen close to the tube and shining brightly, he placed a lead sheet across the screen and its shadow was clear; half the screen was bright where the mystery rays struck and the other half was dark, in the shadow.

It was then that he noticed the skeletal image of his hand and dropped the metal in shock.

X-rays

The unearthly image of his skeleton showed that whatever the mystery rays were, they were indeed like nothing ever seen before or even dreamed of. There could be no doubt of their reality and importance. Yet the cathode ray work was tinged with controversy due to its association with spiritualism, courtesy of Crookes. Crookes had been ridiculed by his colleagues and been denied the presidency of the Royal Society of London as a result of his fantasies. Now Roentgen, who was not a major figure with a reputation behind him, had discovered "X"-rays with this device and seen his mortal remains before their time. The associations were potentially dangerous for him; that was the downside. On the other hand there was no doubt of the singular importance of the discovery. This produced the greater worry, and one common to most who make great discoveries "out of the blue": am I alone in this or is someone else right now aware of it and about to beat me to the glory?

Imagine yourself in Roentgen's situation. You have taken up this work because of Lenard's prior efforts, reported nearly two years before. You have spent 18 months getting prepared, catching up to where Lenard must have been before you can even get started, and then, almost immediately, you discover this bizarre phenomenon. Once seen it is easy to produce. Has Lenard not already seen this? Have others who have been working with cathode rays all missed this? If not, the moment you mention even a hint of it they will all be able to see for themselves, or realize that this is the source of some previously unexplained aspect of their work. Crookes, now dead, had blamed the maker of his photographic plates for their poor quality when he found them fogged; how many others had had the same or similar experiences and would immediately realize the cause and claim priority over you, Roentgen? Instead of being "Roentgen; discoverer of X-rays" you could end up as a footnote on someone else's piece of immortality.

In Roentgen's favour was that no-one had reported rays with such ability to pass through metals as easily as through air, let alone their bizarre ability to cast shadows of bones. So he set about to establish their properties to an extent that he could produce a written paper demonstrating their reality, proving that they differed from cathode rays, and, as such, were a new form of mystery radiation—hence "X" for the unknown.

For the next several weeks he worked and even slept in the laboratory. On the Sunday before Christmas, he took the now famous image of his wife's hand and ring, and six days later wrote up the paper for publication in the *Proceedings of the Physical Medical Society of Wurzburg* where he lived, even though he had not delivered the paper at the December meeting. This is how this "official" news was established. In addition, he produced copies of the shadow images of Bertha's hand and, on New Year's Day, 1896, sent them, along with drafts of his paper, to leading physicists throughout Europe.

One of his colleagues showed them to a visitor from Prague whose father was editor of *Die Presse*, the influential Viennese journal. The news was passed from son to father and *Die Presse* announced Roentgen's discovery on the front page of their 5 January edition. The internet was still a century away, but the sensational news spread hardly any less rapidly than in this modern day. The image of Bertha's hand was probably the first example of what modern public relations term a "photo-opportunity" and *Die Presse* had an international scoop.

The news was an immediate sensation. Scientists around the world quickly reproduced the effects themselves (which is the essential first step that distinguished reality from human delusion) and within a few weeks medics were using X-rays for imaging fractures and as aids in surgery. Roentgen demonstrated the rays to Kaiser Wilhelm and cartoonists suggested that with X-ray vision we could all be seen naked beneath our clothes unless we wore lead-lined corsets.

That X-rays existed and had remarkable properties was immediately apparent, but what X-rays are and how they are produced was a mystery. Today we know that X-rays are a form of electromagnetic radiation—essentially light whose wavelengths are much shorter than those which our eyes record as colours. Cathode rays, by contrast, are beams of electrons, the constituents of atoms that have been torn from their atomic home and turned into electric currents. So X-rays and cathode rays are quite different, though no-one knew that in 1896.

Lenard, for instance, who had discovered cathode rays, jealously believed that the "Roentgen rays" (a common way of referring to X-rays at the time) were just extremely penetrating examples of his own cathode rays or "Lenard rays" (as he might prefer to call them). Lenard bitterly felt that he deserved more credit, that Roentgen had somehow hijacked "his" rays.

In the public dispute of priority there was no doubt that Roentgen's X-ray photographs were the discovery that mattered, to all but Lenard at least. To scientists the primary questions were the nature of X-rays and cathode rays and how they are formed. In the years that followed, X-rays began to shed light, literally, on the deep structure of crystals, leading eventually to that of DNA and the secret of life. X-rays would also lead to the discovery of the worlds within atoms.

From X-rays to DNA

What is the smallest thing that you can see? A dust mote in a sunbeam? A grain of sand? Bacteria? All forms of life on earth have something in common with inanimate matter: they are all made of atoms which are smaller even than bacteria, far too small for our eyes to see. If our unaided naked eye was all that we had, the

Andromeda galaxy and microbes would define the boundaries of our vision. Atoms are beyond normal vision but can be made apparent by other means. Many people may feel that such things are less "real" than those that we can directly perceive. So let's address this problem right away.

What is reality? Answering this superficially simple question is like trying to eat porridge with your fingers. It has occupied the minds of the greatest philosophers such as Ludwig Wittgenstein, whose opinion was that the solution to the mystery of space and time "lies outside space and time." Emmanuel Kant referred to "reality" as the "Ding-an-Sich" or "Thing in Itself" which is intrinsically unknowable and perceived dimly through our natural senses.

We can see, but how we interpret our visions can depend on many factors such as the illumination, the tint of the spectacles through which we are looking, the shadows, and sense of scale set by other objects in the vicinity. The art of the illusionist is based on such vagaries. We can move around and see the "thing in itself" from different vantage points, though the difference between "it" and a hologram is becoming ever harder to detect, at least if sight alone were our only window on the world. To know more we extend our senses by touching the thing: hologram and solid form are immediately distinguished. We can hear, though we are deaf to the octaves where other animals rejoice, such as dogs responding to a high-pitched whistle or bats who avoid crashing to their death by their own high-frequency sonar. Smell and taste add to our subjective perceptions of Kant's "Ding-an-Sich."

One view of the history of science is the building of machines to extend our perceptions. Today these have become extremely sophisticated, such as particle accelerators and huge electronic cameras, whose visions we interpret as recreating the conditions at the heart of stars or even at the start of the universe. But are these "real," the "Ding-an-Sich," or mere delusions? Modern natural philosophers would claim that these are no less real (and no more?) than the subjective perceptions formed with our unaided five senses. They are but the current extremes in a continuum of experience.

Most of us are content with "seeing is believing" as if that is the end of it; if that was the case would molecules and atoms be real? We cannot see them, let alone touch and hear them, yet they are within reach of our natural unaided senses. As I mentioned at the start of Chapter 4, one's nose can detect molecules that are beyond normal vision—which is how the ancient Greeks first came up with the idea that there are fundamental elements to matter. By using our natural senses beyond vision we can access nature that would otherwise have remained invisible. From this it is only a small step to imagine us as cyborgs, extending our senses "artificially" to reach out into space with the telescope as an extended eye or deep into the microcosm with other aids.

X-ray eyes have enabled us to extend our vision to the realm of the atoms. To understand why X-rays can do what our own eyes cannot we first need to think for a moment about what is involved in seeing.

You are seeing these words because light is shining on the page and then being transmitted to your eyes; the general idea is that there is a source of radiation (the light), an object under investigation (the page), and a detector (your eye). Inside that full stop are millions of carbon atoms and you will never be able to see the individual atoms however closely you look with however a powerful magnifying glass. They are smaller than the wavelength of "visible" light and so cannot be resolved in an ordinary magnifying glass or microscope. Light is a form of electromagnetic radiation. Our eyes respond only to a very small part of the whole electromagnetic spectrum; but the whole of the spectrum is alive. Visible light is the strongest radiation given out by earth's nearest star, the sun, and humans have evolved with eyes that register only this particular range. The whole spread of the electromagnetic spectrum is there. I can illustrate this by making an analogy with sound.

Imagine a piano keyboard with a rainbow painted on one octave in the middle. In the case of sound you can hear a whole range of octaves, but light—the rainbow that our eyes can see—is only a single octave in the electromagnetic piano. As you go from red light to blue, the wavelength halves; the wavelength of blue light is half that of red. The electromagnetic spectrum extends further in both directions. Beyond the blue horizon—where we find ultra violet, X-rays, and gamma rays—the wavelength is smaller than in the visible rainbow; by contrast, at longer wavelengths and in the opposite direction, beyond the red, we have infra-red, micro-waves, and radio waves.

Caved dwellers needed to see the dangers from wild animals; they had no need to develop eyes that could see radio stars. It is only since 1945 that we have opened up our vision across the fuller electromagnetic spectrum. The visions that have ensued have been no less dramatic than would a Beethoven symphony have been to a thirteenth-century monk whose experiences ended with Gregorian chants restricted to the single octave of the bass clef.

We can sense the electromagnetic spectrum beyond the rainbow; our eyes cannot see infra-red radiation but the surface of our skin can feel this as heat. Modern heat-sensitive or infra-red cameras can "see" prowlers in the dark by the heat they give off. Bees and some insects can see into the ultraviolet, beyond the blue horizon of our vision, which gives them advantages in the Darwinian competition for survival. It is human genius that has made machines that can extend our vision across the entire electromagnetic range with radio telescopes on the ground and X-ray and infra-red telescopes in satellites complementing optical telescopes on mountain tops and in the Hubble space telescope satellite. We have discovered marvels such as pulsars, quasars, and neutron stars and have flirted with the environs of black holes and a host of other natural wonders that no-one had ever perceived. The visions beyond the rainbow also reveal deep truths about atoms and more.

We can smell molecules but cannot see individual ones normally. A simple demonstration can give you an idea of their size. First, fill a tray with water and sprinkle some flour (or better, if you can get some, lycapodium powder) on the surface so as to make the surface easy to see. Next, take a small droplet of oil and very gently release it onto the surface of the water. It spreads and eventually stops. The reason is that the molecules of oil that were originally all clustered in the spherical droplet have tumbled over each other until they are just one molecule thick. At that point they can go no further; they've reached their natural extent.

You can see the area that they've filled and if you know the volume that you first started with you can compare the volume (a three-dimensional quantity) with the area (two-dimensional) to determine the extent of the one dimension missing. The area divided into the volume gives the height of the molecules, which is revealed to be about a hundred thousandth of a millimetre. This is small, but not so small that it is beyond imagination. One millimetre is easy to see on a ruler, and the marks are probably about one-tenth of that. These lines in turn will easily accommodate more than 10 human hairs laid side by side, and so you have an idea of one-hundredth of a millimetre. The number of such molecules that would form the diameter of a hair is around 10,000— the numbers of people that might attend a minor league football game. So individual molecules are beyond normal vision, but not so far away.

Oil molecules are quite long, with tens of atoms loosely clustered together. If individual atoms were laid end to end, up to a million would cover the thickness of a hair. It takes some imagination but a million is possible to comprehend. Visualize the crowd at a World Cup soccer match and multiply it 10 times; now imagine that number in the width of a hair. Atoms are probably at the limits of our imagination.

Our inability to see atoms has to do with the facts that light acts like a wave and waves do not scatter easily from small objects. To see things, the wavelength of the beam must be smaller than the thing you're looking at, and therefore to see molecules or atoms needs illuminations whose wavelengths are similar to or smaller than them. To gain a feeling for how big a task this is, how far away they are from sight, imagine the world scaled up 10 million times. A single wavelength of blue light, magnified 10 million times, would be bigger than a human, whereas an atom on this scale would extend only 1 millimetre, far too little to disturb the long blue wave. To have any chance of seeing molecules and atoms needs light with wavelengths that are much shorter than these. We have to go far beyond the blue horizon to wavelengths in the X-ray region and beyond.

X-rays are light with such short wavelengths that they can be scattered by regular structures on the molecular scale, such as are found in crystals. The wavelength of X-rays is larger than the size of individual atoms, so the atoms are still invisible. However, the distance between adjacent planes in the regular matrix within crystals is similar to the X-ray wavelength and so X-rays begin to discern the relative positions of things within crystals. This is known as "X-ray crystallography" which has become the main and revolutionary use of X-rays in science.

An analogy with the hieroglyphics of X-ray crystallography is if one thinks for a moment of water waves rather than electromagnetic ones. Drop a stone into still water and ripples will spread out in circles. If you were shown a picture of these circular patterns you could infer where the stone had been. A collection of synchronized swimmers diving in would create a symmetrical series of circles that build up into localized splashes, where two peaks meet and cancel out where peak meets trough. From the resulting pattern one could, with more difficulty, deduce where the swimmers had entered the pool. X-ray crystallography involves detecting the multiple scattered waves from the regular layers in the crystal and then decoding the pattern to deduce the crystalline structure.

One can take pictures of very complicated molecules, such as DNA, which is the most famous application of X-ray crystallography. The sequence of events that brought us from Roentgen's X-rays to the discovery of the double helix of the DNA molecule and modern genetics is an ideal example of how science works— of how you have no idea of what developments will turn up in the future as a result of fundamental discoveries being made today.

Imagine 1894: the discovery of X-rays is still a year away. People at that time thought they were almost at the threshold of an explanation of everything and then suddenly X-rays appear. Initially they are a mystery, and then X-ray crystallography develops. This is the first step in the scientific and technological process. A newly discovered phenomenon becomes a tool that is put to use in science. With this tool—an extension of our senses in effect—further marvels come into our range. The molecular structures of salts, proteins, and DNA are solved.

The moment the double helix structure of DNA was discovered, reproduction, genetics, the nature of life itself became open to investigation. A further 40 years has brought the genetic revolution, biotechnology, and eradication of diseases. No one had any idea a hundred years ago what would come from Roentgen's X-rays. X-rays have revealed the mirror asymmetric profundity of the DNA helix. The speculations of van't Hoff on left- and right-handed molecules, the handedness of amino acids, and the glorious code of the DNA spirals are, through X-ray eyes, revealed as "real." A thousand years in the future, much of our modern view of the universe will appear naive, however the essential truth about the CGAT spiral of DNA will surely be forever. Crick and Watson's discovery is widely regarded as the most profound of the twentieth century and we can be as certain as we can of anything that their interpretation of DNA is an all-time truth.

Chapter 6

"Electrick virtue"

Nature has buried its secrets deep but has not entirely hidden them. Clues to the restless agitation within its atomic architecture are all around us: the radioactivity of natural rocks, the static electricity that is released when glass is rubbed by fur, the magnetism within a lodestone, sparks in the air, and lightning are just a few of them. Great discoveries are popularly believed to be the preserve of genius, whereas more often the critical feature is being in the right place at the right time and with the right preparation. The response of Henri Becquerel in France to the new X-rays is a prime example.

The mysterious rays were being put to use even though none yet knew what they were. While the medics saw the bone shadows and immediately applied this property to their own primary interests, many physicists repeated Roentgen's experiments and tried to extend them in the hope that X-rays would somehow reveal new secrets. For Becquerel the images stirred a long-time memory such that within weeks he had set science onto another course.

Becquerel's grandfather had been intrigued by phosphorescence and had wondered why some metals and minerals glow in the dark. Henri's father, Edmund, also had investigated this phenomenon in experiments involving uranium compounds, and Henri had helped him. The vague ideas on the nature of chemicals that give out light in the cold were that the atoms were somehow soaking up bright daylight and then releasing it gradually as a dull glow. This is indeed the case in some cases, but not all. No one suspected that in the case of uranium it was the atoms themselves that were pouring out energy relentlessly and had been doing so for millions of years. One reason for this misconception was that those who believed in the existence of atoms (and that was by no means everyone in 1896) imagined them to be hard, impenetrable spheres rather than the highly structured objects that we now know. The answers could hardly have been guessed as in 1896 they were still three discoveries removed from the required state of knowledge. Within 10 years all this would change as a result of what Henri Becquerel was about to discover.

Becquerel had learned of Roentgen's X-rays in the first week of January 1896. By this time he was 44 years old, had a strong track record in research into phosphorescence, uranium compounds, and photography, and was a member of the French Academie de Sciences. The news of Roentgen's discovery and the startling photographic images immediately resonated with Becquerel's experiences and made him wonder if X-rays and phosphorescence were related.

Once he had had the idea, it was obvious how to test it. First, he exposed some phosphorescent crystals to sunlight for several hours so that they were energized as usual. Then he wrapped them in opaque paper, placed the package on top of photographic emulsion, and put them all in a dark drawer. If the fluorescence emitted only visible light, none would get through the opaque paper to reach the photographic emulsion, whereas any X-rays emitted by the crystal would pass uninterrupted to the photographic plates and fog them. As an extra test he placed some metal pieces between the package and the photographic material so that even X-rays would be blocked and leave a silhouette of the metal in the resulting image.

When he developed the plates he found that they had indeed been exposed and, most important, contained shadow images of the metal plates. He reported the discovery to the Academy of Sciences of 24 February 1896, only six weeks after having first learned of X-rays. He had proved without doubt that the crystals were responsible. Also he had noticed that uranium compounds were particularly good performers in his experiments and commented on this to the members of the academy. But he was wrong in his belief that it was his exposure of them to sunlight that had provided the energy that set the process in train.

The true secret was still to be revealed. It was during the following week, as he continued his experiments, that he made his major discovery: uranium radiates energy spontaneously without need of prior stimulation such as sunlight. In fact, sunlight has nothing to do with it as serendipity was about to show.

The last week of February in Paris was overcast. Mistakenly believing that sunlight was needed to start the effect, Becquerel saw no point in continuing until the weather improved. As luck would have it, it remained cloudy all week and by 1 March he was becoming frustrated. At this point, for want of something to do, he decided to develop the plates anyway expecting, as he later reported, to find a feeble image at best. Oh lucky man! To his great surprise the images were as sharp as before showing that the activation took place in the dark.

Unlike Crookes, who years before had missed discovering X-rays by his mistaken belief that fogged photographic plates were due to faulty manufacture, Becquerel pursued the new phenomenon. Perhaps his inspiration came from the association with the uranium that he had already commented on to the Academy. In contrast to his earlier experiments, the uranium in the drawer had not been exposed to bright light prior to the photographic process, which suggested either that it reacted even to dim light or, amazingly, that it was emitting penetrating radiation of its own accord. By repeating a similar set of experiments to before, but this time with completely unexposed uranium, he soon established the world shattering result that energetic, penetrating radiation could come from inert matter spontaneously and without need of prior stimulation by light.

Although this discovery of spontaneous "radioactivity" is today recognized as seminal, it is ironic that at the time it made no special impact. Radiations were the novelty of the decade: cathode rays, X-rays, Lenard rays or Roentgen rays, radio waves, along with light emitted from cold crystals or even from living creatures such as fireflies were all vying for attention, and so Becquerel rays were originally regarded as merely one more for the list. The real birth of the radioactive era was when the Curies discovered radioactivity in other elements, in particular radium, so powerful that it glowed in the dark. (It was Marie Curie who invented the term "radioactivity.") By 1903 the significance was fully realized and it was appropriate that the Curies shared the Nobel Prize with Becquerel.

It was the pursuit of radioactivity that led to the discovery of the nuclear atom in the early years of the twentieth century. The revelation that the atom has an inner structure built on the mutual attraction of opposites, in the form of negative and positive electrical charges, will begin to bring us to the deep symmetry at the heart of the matter. The first step on this journey involves the discovery of the electron.

The ubiquitous electron

What a difference a hundred years make. Stanley voyaged for two years to talk with Livingstone on behalf of his newspaper; today he would call up on his mobile phone and spend the next 729 days on other tasks. The communications revolution is effectively shrinking the world, as contact is instant via satellite phone, without need of travel, and the pace of life has accelerated. These are some of the results that have followed from 30 April 1897, when Joseph John Thomson marched into the Royal Institution in London and announced that he had discovered the electron, a fundamental constituent of the atom.

Electrons flow through wires as electric current and power industrial society; they travel through the labyrinths of our central nervous system and maintain our consciousness; and their motions from one atom to another underpin chemistry, biology, and life. It has been estimated that modern electronics increase the GDP of the world by the equivalent of a million pounds every minute, a trillion dollars each year. The VAT and sales tax alone on this comes to a larger amount than world governments spend on fundamental scientific research. Thomson's electron is rightly regarded as the farthest reaching of the great trinity of discoveries (that includes X-rays and radioactivity) and the centenary of its discovery was recently celebrated. Germany and France celebrated their centenaries with Becquerel adorning a 500-franc bank note and Roentgen, a stamp, such that the world has been made aware of their justified pride. (What irony that Britain, despite the urgings of many prominent scientists and industrialists, chose to celebrate on their stamps not the centenary of Thomson's discovery of the electron but instead the 1897 birth of the childrens' author, Enid Blyton.)

The electron is the lightest particle with electric charge; it is stable and ubiquitous. Our limbs, bodily features, and indeed the shapes of all solid structures are dictated by the electrons gyrating at the periphery of atoms. Electrons are in everything and it is quite easy to liberate them: during televised lectures in London's Royal Institution at Christmas 1993, I formed a human battery with three children and made electric current from our bodily electrons. Electrons are involved in many nuclear reactions such as those that power the sun. Gravity rules the universe but it is electromagnetic forces, and their agents, the electrons, that give shape, form, and structure, especially here on Earth. The electron is present in all space and for all time: modern theory suggests that the electron was the first material inhabitant of the universe in the act of the Creation and if the universe expands forever, electrons will probably be among its last remnants when the lights go out.

It is surprising that the discovery of the electron, which was there for the finding by any of several scientists, fell to Thomson. He had won a scholarship to Cambridge in 1876 to study mathematics. This led to a fellowship at Trinity College where he continued in his primary strength of mathematical physics. In 1884, when the professorship at the Cavendish Laboratory fell vacant, the authorities wanted to hire Lord Kelvin, universally acknowledged as the leading experimentalist of the nation, but he preferred to stay in Glasgow. This was the third occasion that Cambridge had tried, and failed, to attract him. Thomson was a surprise alternative choice, though it is hard to imagine that Kelvin could have achieved greater things than Thomson would during his subsequent 50 years at the helm.

It was soon after Thomson took up the post that he began investigating the nature of electricity. As a child, I bought a book, "Questions and Answers in science" in a jumble sale. "What is electricity?" it asked. "Electricity is an imponderable fluid whose like is a mystery to man" it opined in Victorian melodrama. The book was published in 1898—the year after Thomson's discovery; news travels faster these days.

"Electrick virtue"

Electrical phenomena have been known for thousands of years and some of the special properties had been investigated by the Ancient Greeks. Pull a plastic comb through your hair, or rub your coat with amber, and the plastic or amber can pick up pieces of paper. On a dry day a rapid combing may even induce sparks. Glass, jet (compressed coal), and gems also have this magical property of clinging to things after rubbing. By the Middle Ages the courts of Europe knew that this weird attraction is shared by many substances. The magic was that substances only gained the "power" after rubbing and from this William Gilbert, court physician to Elizabeth I, decided that electricity is some "imponderable fluid" (as in my 1898 book) that can be transferred from one substance to another by rubbing. This "electrick virtue," as he referred to it (or electricity in more modern parlance) takes its name from the Greek word for amber, "electron."

The next step was the discovery that electrified glass can transfer its "electrick virtue" to other bodies, either by touching them or by an intermediary wire: the idea of electric current took hold. Then they found that when electricity was transferred from glass to two separate pieces of metal, the latter would repel each other. Previously "electrick virtue" had attracted things, as in its ability to pick up pieces of paper; now it could also repel. From this grew the idea that there are two kinds of electricity, which became known as "vitreous" and "resinous."

Unification (combining two or more ideas into one) is a driving force within physical science. As Sherlock Holmes would remind Watson: when two strange things happen they are likely to be related. The recognition that there is really only one "electrick virtue" manifesting itself in different ways came from Benjamin Franklin. A versatile man, later to be one of the signatories of the American Declaration of Independence, Franklin became fascinated by electricity in the 1740s after seeing a popular demonstration of electric phenomena in Boston, New England, by a visiting lecturer from Britain. He began his own experiments, including the famous flying of a kite in a thunderstorm (though it is unclear whether he actually performed this personally), attempting to catch the electric currents associated with lightning.

A lightning cloud is really just a natural electrostatic generator, but one capable of making millions of volts and sparks that can kill. Franklin's insight was that bodies contain the electric power latent within, that it can be **transferred** from one body to another.

Fig. 6.1 Benjamin Franklin on $100 note from USA.

Gaining some of the "electrick virtue" is what Franklin defined as being **positive**; losing it, **negative**—the total amount of "electrick virtue" or "charge" is preserved. When two bodies are rubbed together, be they glass and silk or amber and fur, the amount of positive electric charge on one is equal to the amount of negative on the other; the total charge is zero throughout. Today, following Thomson, we would say that atoms contain electrons and that it is electrons being transferred from one body to another that carry the electric charge. One hundred and fifty years elapsed between Franklin's insight and Thomson's discovery, by which time the convention of "positive" and "negative" had become irrevocably established. So, by historical accident, we have inherited the definition that electrons carry negative electric charge.

The total negative charge is equivalent to the total number of electrons added. Conversely, the total positive charge gained equals the total number of electrons removed. When one rubs cold solid objects like glass and silk together, electrons are transferred—not created or destroyed. Today, two hundred and fifty years later, this conservation of electric charge is presenting us with one of the great mysteries of existence, but Franklin could not foresee that any more than could Thomson. More of that later.

That electrons transfer from amber to Teflon and not the reverse is an empirical fact and the result of the atomic structure at the surface of these materials. This is a complex area of research and even today it is a computational challenge to get from known atomic physics to predict which way electrons will go. The flow of electrons and the bonding of materials by electrical attractions, such as in the development of new adhesives, is as much an art as science. The underlying physical principles are understood but computing the implications is at the limits of ability.

We know today that electrons contribute less than 1 part in 2,000 of the mass of a typical atom and, as only a small percentage of them are involved in electric current anyway, the change in mass of a body when electrically charged is so trifling as to be undetectable. How then was the "imponderable fluid" or "electrick virtue" to be isolated, catalogued, and studied? Electricity normally travels in something, be it wires, gases, or even through our bodies. As it was impossible to "look" inside wires, the idea developed that if you could get the wires out of the way and study the sparks, then maybe there would be a chance of seeing what electricity consisted of. Indeed, lightning showed that electric currents can pass through the air and from this grew the idea that the flow of electric current might be revealed "out in the open," away from the metal wires that more usually conduct it and hide it.

So scientists set about making sparks in tubes and removing more and more gas from the tubes to see if only the electric current would remain. It turned out though that air at normal pressure obscured the flow of the electrons and it was only after most of the air was removed that the bizarre presence became visible. First, at one-fiftieth of normal atmospheric pressure lights began to glow in the gas (which is where the story of Chapter 5 began). The colours of the lights depend on the gas, the most familiar being the yellow light of sodium and the green of mercury vapour common in modern illuminations. They are caused by the current of electrons bumping into the atoms of the gas and liberating energy from them as light. As the gas pressure dropped further, the lights disappeared but a green glow developed on the glass surface near to the cathode (the source of the current). This effect occurred independent of the position of the anode (the metal plate attached to the positive terminal of the electric circuit) and suggested that something was coming out of the cathode.

The critical discovery in 1869 that objects inside the tube cast shadows in the green glow proved that the rays were in motion, coming from the cathode and hitting the glass, except when things were in the way. The British scientist Sir William Crookes, who began studying cathode rays in 1879, found that a magnet would deflect them, which suggested that they consisted of negatively charged particles. It was around this time that Crookes became obsessed with the notion that he was producing ectoplasm. The challenge of determining what these mystery rays *really* consisted of moved elsewhere.

Thomson and the electron

By 1890 two camps had developed. Most French and British scientists, influenced by Crookes, believed that the rays consisted of electrically charged particles. Opinion in Germany differed. Scientists there were influenced by Heinrich Hertz who, in 1883, reported that the rays were not deflected by electric forces. This led to the local received wisdom that the rays travelled in straight lines, were unaffected by gravity, and so were waves like light.

Hertz discovered electromagnetic waves propagating across space in 1885 as a by-product of his earlier wrong ideas. The reason for his incorrect interpretation of the cathode rays was that the electric fields available to him were too feeble to deflect the fast-moving electrons measurably. Pursuing this false belief did bring him two years later to the real electromagnetic waves. Meanwhile the quest to determine the nature of cathode rays continued and stimulated Roentgen, leading to his discovery of X-rays (a secondary phenomenon when the cathode rays hit metal in the tube). In turn, this had brought Becquerel to stumble on radioactivity (which has no immediate connection to cathode rays at all). Throughout all of these events the nature of the cathode rays remained unresolved and was partly the reason why the nature of X-rays was initially so uncertain.

In 1895, Jean Perrin in France found that cathode rays deposit negative electric charge on a collector inside the tube. Whether this was because the cathode rays were negatively charged or because negative charges were being ejected from atoms in the residual gas was moot. Thomson succeeded were Hertz had failed and showed that it was indeed the cathode rays that carried negative charge; Thomson's advantage was in being able to reach lower pressures (which enabled the electrons to flow more easily) and more powerful electric fields (which deflected the beams more). Thomson could tell where none had before that the cathode rays were repelled by the negative electric plates and attracted by the positive charged plates; like charges repel and unlike attract is the rule and from this Thomson proved conclusively that the constituents of cathode rays, the electrons, are negatively charged.

This resolved any residual caveats about the source of Perrin's negative charge, but raises the question: in what sense was Thomson the discoverer of the electron? Quantitative rather than qualitative measures enabled him to make a generalization or an intellectual leap or an inspired guess—depending on how you interpret it. In any event, he was correct.

Thomson's key ingredient was in using both electric **and magnetic** fields to move the beam around. When the beam hits the glass at the end of the tube (essentially the "screen" in the modern television tube of which Thomson's device was a prototype) it makes a small green spot. Thomson used electric forces, by connecting the terminals of a battery to two metal plates inside the tube. One plate was charged positive and the other negative; the beam was attracted up by the positive one and repelled by the negative one. This proved that the beam definitely was electricity. He surrounded the tube with coils of wire which create a magnetic field; this too deflects the beam. Then he did a clever thing. He used the electric and magnetic forces together to cancel each other out and return the green spot to its original position. This might seem like a strange thing to do, but electrostatic forces deal with electric charges while magnetic forces are concerned with moving electricity and the mathematics of the experiment reveal that when you cancel these two out, you can calculate the velocity and the mass per unit electric charge of whatever is moving. By this means Thomson deduced the properties of the constituents of the electric current.

He measured these in a series of experiments using a variety of gases in the tube, different metals in the cathode, and for a range of velocities for the cathode rays. He found that in each and every case the ratio was within a factor of two of 10^8 Coulombs* per gram. Such a large number each time convinced him that this result was a property of the rays and independent of the gas and cathode materials.

This is already an important advance, but we are still short of answering the critical question: where is the electron to be seen in all this? This is not at all obvious; witness the story of Walter Kaufmann. Who is he you may ask? His name is nowhere as well-known as Thomson, he won no Nobel Prize, he is not among the chosen few who are credited with the seminal discoveries of our culture, yet he did essentially the same experiment as Thomson, at the same time as Thomson, and his results were better than Thomson's! We know this because today, with modern instruments, it is possible to measure the ratio of charge to mass to an accuracy of 1 part in 10,000. Compared with this Thomson's value was no better than within a factor of two, whereas Kaufmann, in 1897, was some 50 times more accurate than Thomson. Nonetheless, it is Thomson who is recognized as discoverer of the universal electron as Kaufmann, with mental "myopia" from his belief in the German school that cathode rays were a form of electromagnetic wave, failed to realize the full implications of his result.

Kaufmann, in Germany, was unprepared and lost. By contrast, Thomson in Britain was doubly fortunate. First, the opinion there was that cathode rays were indeed electrically charged. Second, he knew of some work by Pieter Zeeman, the Dutch spectroscopist. Zeeman had been studying the effect of magnetic fields on the light emitted by atoms of sodium. You can do this by adding some element such as sodium to a flame and passing the light through a prism or a diffraction grating to break the light up into its component colours. These will include a series of bright lines which in the case of sodium include two particularly intense yellow–orange ones that are the source of the familiar colour of sodium street

---

*A coulomb is a unit of electric charge. It is defined as the amount of charge transported by a current of one ampere in one second.

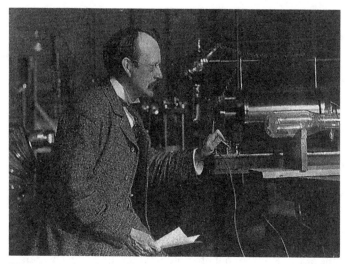

Fig. 6.2 J. J. Thomson and his cathode ray tube.

lamps. These lines are very sharp but Zeeman noticed that they broadened in a magnetic field; the stronger the magnetic field, the broader the lines became. Theorists realized that with this knowledge they could determine the ratio of the charge and mass of the objects that were the electrical source of the light. As a result, a value for the charge to mass ratio of the carriers of electric currents within atoms had been computed by 1896, a year before Thomson did his work. Thomson's value in 1897 was near enough to this to give him the courage to make the leap.

Zeeman's work had shown that there is something that carries electric current *in* atoms; Thomson had shown that this same something has a real existence *outside* atoms. The inference is that the objects carrying the current in cathode rays are electrically charged constituents *of* atoms. Thomson made his bold assertion that they were common to all elements based on the fact that the properties of his electrons were always the same when they were emitted as cathode rays from a range of metals, and furthermore were present within sodium and other elements courtesy of Zeeman's effect.

The electronic atom

An essential truth, once stumbled upon, rapidly ties up loose ends in other phenomena and gains acceptance as "reality." Thomson's electron is a prime example. It was already known that the ions that carry electric current in water, as when salt (sodium chloride) splits into positively charged sodium ions and negatively charged ions of chlorine, have charge to mass ratio that vary from element to element. However, after Thomson had measured this ratio for the cathode ray "electrons," it was immediately apparent that the electron value was more than a thousand times different from any of those for atomic ions and moreover, the same every time, independent of what metal the cathode was made of. The reason was quickly understood: ions are the result of knocking one or two electrons out of an atom (as for the positively charged sodium) or of adding electrons (as in negatively charged chlorine). The mass of an electron is trifling on the scale of an atom's mass; hence the huge value of charge to mass ratio for an electron (in cathode rays) relative to an ion (where the full mass of the atom appears in the denominator of the ratio). This added to confidence in Thomson's hypothesis that lightweight electrons are pieces of atoms.

Once it was realized that these constituents of atoms—the pieces of electrical current—are at least 2,000 times lighter than the smallest atom, scientists understood the enigma of how electricity would flow so easily through copper wires. The existence of the electron overthrew forever the age-old picture of atoms as the ultimate particles and revealed instead that atoms have a complex structure.

Atoms consist of a compact centre (the nucleus) which is positively charged, surrounded remotely by the flighty electrons that are negatively charged. Although overall atoms may have no electrical charge, they do contain electrical charges and intense electric fields within. It is these that give structure to matter: it is gravity that pulls us to the ground but it is the intense electrical activity within our constituent atoms and those of the ground that stop us sinking to the centre of the earth.

It is these electrical forces, caring naught for left and right, that glue neighbouring atoms to one another, forming the more complex molecules of life. Carbon atoms have their electrons whirling around in such a way that four of them are free to link with neighbouring atoms. It is this quartet that gives carbon its four "tentacles" (page 61) with which neighbouring atoms are ensnared. Each element has its own special number of such electrons, or tentacles, and it is matching these links that determines how different elements combine. For example, hydrogen has but one; oxygen has eight electrons in all but at any time only two are available for linkage—hence the affinity of two hydrogen for one oxygen making H_2O, as in water.

As we saw in Chapter 4, it is the four tentacles of carbon that give it its rich and tantalizing complex molecular structures. We saw also how easy it is for asymmetric structures to arise and hence the emergence of molecular structures that are mirror images of one another. The existence of the world of mirror molecules was first revealed by Pasteur's discovery of the rotation of plane-polarized light (page 67). We can now see why this occurred. The molecular glue consists of electric and magnetic forces; light, meanwhile, consists of electric and magnetic fields that are oscillating. When these oscillating fields encounter those gluing the molecules, their oscillations will be disturbed; the electric and magnetic fields of the departing radiation will be skewed relative to how they were before.

The electric and magnetic fields in a left-handed molecule will have some complicated form; those in the right-handed version will have the same complicated form, except that they will all run the opposite way. The attractions and repulsions felt by a beam of polarized light will also be reversed. Whichever way the one molecule skews the light beam, its mirror version will skew in the opposite sense. Where electric and magnetic fields deflect to the left in one case, they will deflect in the same way but to the right in the other. The deep-seated mirror asymmetry of life's molecules is thus revealed by the opposite twisting of polarized light.

Chapter 7

The heart
of the matter

In October 1895, one month before Roentgen discovered X-rays and two months before the dramatic announcement of his discovery, a young New Zealander, Ernest Rutherford, left home and travelled halfway around the world to England and the Cavendish Laboratory in Cambridge. His curriculum vitae in summary runs: identified two forms of radioactivity in Cambridge, moved to Canada and discovered transmutation of the elements for which he won the Nobel Prize, then to Manchester where he established the nuclear atom, and finally returned to Cambridge, succeeding J. J. Thomson as Cavendish Professor. By then Rutherford had already made discoveries sufficient for three lifetimes and was the greatest experimental physicist of the day.

Thomson is renowned for his discovery of the electron, in April 1897—a monumental event that has transformed twentieth-century technology and was the culmination of years studying the conduction of electric currents through gases. When Rutherford first arrived in Cambridge in the autumn of 1895, Thomson was already deep into this research, his excitement intensifying when he learned that Roentgen's mystery rays ionized gases, liberating electric charge from their atoms. A compelling new field was opening up, Thomson was near its frontier, and a brilliant young research student had arrived fresh in his laboratory. It was natural for Thomson to guide Rutherford into this new field.

It is tempting to think that the rest, as they say, is history, and that this explains how Rutherford began his career that revealed to the world the nuclear atom. The actual course of events was more roundabout and in part the result of wrong advice.

Rutherford was no ordinary novice. As a student in New Zealand he had been tinkering with magnets and electrical equipment and had discovered how to transmit and receive radio signals; indeed, in 1895, when he came to Cambridge, he was more expert than Marconi at that time. Naturally Rutherford was planning to research and develop the reception of electromagnetic waves. This created a dilemma for Thomson, and so before deciding whether to divert Rutherford into the newborn field of X-rays, he sought advice from Lord Kelvin on the commercial possibilities of radio. If the great man had been more enthusiastic, history could have been different: Rutherford may have become famous for radio and someone else become known as the founder of the atomic age. However, Kelvin was sceptical about the future of radio and so Thomson directed Rutherford to investigate the new rays, leaving the field of radio to Marconi (who otherwise might have been a footnote to the "Rutherford Wireless and Telegraph Company").

Deep within matter, atoms are announcing their presence by spitting out rays into the air. For billions of years they have done so, forming the elements of life, maintaining the inner heat of the earth, awaiting discovery by science. At last, as winter turned to spring in 1896, Henri Becquerel chances upon them, not knowing what they are nor from where they originate. While Marie Curie was trying to determine what elements other than uranium are sources of the rays, and while Becquerel was continuing to improve his original measurements involving uranium, Rutherford initially used the rays as a tool, a convenient source for ionizing gases. This was but a brief interlude as he quickly realized that the greater question was the nature of the emanations themselves. He turned the focus of his enquiry on its head and used the ionization of gases as a means of studying radioactivity, rather than the other way round.

One way of doing this is to use a gold leaf electroscope whose basic principle is simple. First you rub cat's fur on a glass rod. This sounds like black magic but it has a real purpose: it makes the rod electrically charged. Then you touch a metal plate with the rod, which transfers the charge to the metal. At the end of the metal are two thin pieces of gold foil that move apart by electrical repulsion as the electric charge runs down the rod onto them. The simple apparatus is then ready for use. Left as it is in dry air the gold leaves stay apart in a "V" shape, but if the air around them becomes ionized, the charge leaks away from the pieces of foil. The electrical force that previously kept the foils apart is reduced and the leaves collapse towards one another. The faster the leaves fall, the greater is the leakage, the more ionization, and, the crux, the stronger the radiation must be.

As I have described it sounds rather rough and ready. The actual apparatus was more precise than this, though the principle underpinning it was the same. Nonetheless, this is an example of Rutherford's genius: the ability to use the simplest of tools to tease out the most profound truths.

With this he was able to see how easily the radiation could be absorbed by various materials. He covered the uranium with thin sheets of aluminum foil and found that the intensity of the radiation dropped as he expected. As he added more and more aluminum, the radiation continued to die off, though less slowly than before. Whereas initially a very thin piece of foil had been sufficient to cut back the intensity by half, he found that several millimetres of foil were required to absorb the remaining radiation. He deduced that there are two types of radiation: one (which he called "alpha") being easily absorbed and the other ("beta") being more penetrating.

When subsequently a third type was discovered, distinct from the previous two, it was natural that following Rutherford's classical "alpha" and "beta," this new variety became known as "gamma" radiation. The gamma rays are a form of high-energy X-rays; while X-rays may be emitted from electrons bound deep in large atoms, the gamma rays are emitted from within the atomic nucleus. Rutherford himself would later show that the beta rays are electrons (or as he put it at the time "similar to cathode rays," which J. J. Thomson in 1897 had identified as electrons), and that alpha rays consist of particles that are the nuclei of helium atoms. It took Rutherford 10 years to solve completely what was going on, and we shall come to that in due course.

What had grabbed Rutherford's attention more immediately was that in the course of identifying the two forms of radiation, he had noticed that the radioactive element thorium was emitting a gas that was itself radioactive. This was 1900; Pierre and Marie Curie had already found in 1898 that radium also gave off a radioactive gas that was similar, probably identical, to what Rutherford was now finding—but what was it? Identifying this gas was imperative. Rutherford by now had moved to McGill University in Montreal and met Frederick Soddy of the chemistry department who took up the challenge.

Alchemy

Soddy found that the gas was chemically inert: it was argon, another element! This was a complete surprise. Radium and thorium, two elements, supposedly permanent unchanging foundation stones of the material universe, were in fact spontaneously able to emit atoms of argon, another of the supposedly fundamental elements. In their observation of spontaneous disintegration of the elements, Rutherford and Soddy had overthrown the paradigm that the atomic elements are permanent and perfect. In so doing they had made one of the greatest discoveries and turned alchemy into a science. Next they investigated how uranium, radium, and thorium changed their elemental form as they emitted alpha and beta radiation. The "new" elements that were created turned out to be a rag-bag of complexity; Rutherford and Soddy's disentangling of the elemental code was a *tour de force*.

Their first breakthrough was the discovery that each radioactive product has a characteristic half-life—that is the amount of time that it takes for the intensity of the radiation to halve, or equivalently for half of the atoms to transform into atoms of a new element. The half-life of uranium is 4.5 billion years, that of radium 1,600 years, while one of thorium's decay products has a half-life of only 22 minutes.

Soddy the chemist was able to show that the decay products consisted not simply of a variety of elements but even of physically varient forms of the same elements. He named these physically different but chemically identical forms "isotopes," from the Greek for equally placed ("placed" in the periodic table of elements). This happens because atomic nuclei are built from clusters of two types of particles—"protons" and "neutrons." The amount of positive electrical charge on the nucleus (carried by "protons") attracts a counterbalancing number of negatively charged electrons and it is this number in an electrically neutral atom that makes the chemical properties by which we recognize that element. Nature may add a number of neutral "neutrons" in the nucleus, adding mass but no electric charge and thereby leaving the chemistry of the resulting atom unaffected. The masses of isotopes differ because of their greater or fewer numbers of neutrons.

The chemistry remains the same but the physical properties can be changed dramatically. Different isotopes of a single element may have very different half-lives. For example, the nucleus of the light element beryllium when made from four protons and five neutrons is stable; however, if made from four protons accompanied by only four neutrons, it fragments into two alpha particles in about a millionth of a billionth of a second, holding the record for the shortest measured half-life of all nuclei. The huge range of values for the half-lives enabled the detection and classification of the products, or "transmuted elements," and it was this that enabled Soddy to do his work. They are also useful in forensic science and in tracing the origin of life, of which more later.

In parallel, Rutherford was studying the alpha and beta radiations. He conclusively proved that the beta rays consisted of electrons and strongly suspected that the alpha were positively charged helium atoms. The proof of this suspicion would not arrive until 1908, to be announced at the Nobel ceremony that he would share with Soddy for their work on transmutation.

Marie Curie

Marie Sklodowska was born in 1867 in Poland and emigrated to Paris in 1891 with no more than £15 in her pocket. Even after allowing for the change value of money, this was an act of courage for a woman wanting to become a scientist in those days. She met Pierre Curie, who was eight years her senior and already on the faculty of the Ecole de Physique et de Chemie Industrielle. His early work had been on the symmetries of crystal structures and on piezoelectricity (the ability of certain crystals that do not have a centre of structural symmetry to become electrically polarized when subject to pressure). They were married in July 1895 after Marie had passed her qualifying exams. Their daughter Irene (a future Nobel laureate) was born on 12 September 1897, following which Marie began her Ph.D. project. Pierre suggested that she investigate the new "Becquerel radiation."

First, she repeated Becquerel's work. He had found that the radiation ionized air and had used an electroscope to investigate it. Instead, Curie used an electrometer with piezoelectric quartz as an essential piece; this input was an innovation from Pierre and proved important in enabling her to make quantitative measurements of the radiation more accurately than had Becquerel and at a similar level of detail to what Rutherford was then doing in England. She found that the intensity of radiation is proportional to the amount of uranium in the compound and independent of its chemical form, whether oxide, salt, or uranium metal.

This confirmed that the radiation is an atomic property of the element uranium, which is what Becquerel had suspected.

Whereas he had concentrated on studying the radiation from uranium in the hope of learning more about it, Marie Curie took off in a different direction to see whether any other elements showed the radiation. She soon found that thorium, a silvery white metal found in minerals, also does. This proved that the mystery radiation was more than a mere curiosity of uranium but was a general natural phenomenon and she gave it the name that we all know it by today: "radioactivity."

It was at this point that she made her lucky, or inspired, leap. Instead of continuing to examine individual elements and compounds, she turned her attention to natural ores. She found that minerals containing uranium and thorium are radioactive, as of course they should be, but the surprise was that the radioactive intensity of some of them was greater than could be accounted for by their uranium and thorium content alone. For example, the uranium ore known as pitchblende is very radioactive when dug from the ground. Marie Curie was able to reproduce its radioactivity with pure substances out of bottles and found that the radioactivity exceeded that from the uranium; the ore must contain some additional impurity, some element, that is highly radioactive.

To extract this mystery element the Curies developed a new science: radiochemistry. The only thing known about the element was that it is radioactive. Their strategy was to take the ore, dissolve it if possible, separate its components by standard chemical analysis, and then see where the radioactivity ended up. By then selecting the highly radioactive extract, the concentration of the new element was increased. They found the culprit remarkably quickly. Having begun the search only at the end of 1897, by April 1898 she had determined which elements are radioactive and by July had isolated the previously unknown element; in honour of her birthplace it was named "polonium."

Then followed a more dramatic discovery. Continuing the purification process, by September 1898 they discovered a further hitherto unknown element whose radioactivity is so powerful that in pure form it glows in the dark and is warm to touch: radium.

Rutherford and Soddy during their investigations of radioactivity were naturally impressed by the immense power of radium and measured the energy given off during its disintegration. They found that a single gram emitted at least 100 million calories, maybe even a hundred times more than this, and announced that the amount of energy emitted in radioactive change was "at least 20,000 times and may be a million times" greater than the energies emitted for the same weight in chemical reactions. Here was the first hint of the awesome, and awful, potential in the new discoveries about the atom.

In 1903 Rutherford visited the Curies in Paris. After dinner they retired to the garden where it was dark; Marie Curie then brought out a test-tube filled with zinc sulphide containing some dissolved radium. It glowed brilliantly, making visible the energy emitted as the radium died, initiating a tumble down the periodic table towards lead, the end point of the radiation chain. The illumination was bright enough to show how inflamed Pierre Curie's hand had become from the continuous exposure to radiation; here was an early example of its hazards.

Some years ago I was privileged to be allowed to study Rutherford's notebooks which are housed in the Cambridge University library. They have been carefully catalogued and stored for nearly 80 years. While turning over one page I noticed a piece of paper trapped in the spine that looked as if it had never been opened. The writing on the outside was the answer to an exam question that some long forgotten student must have written, but what was it doing stuffed in the great man's notes? Anticipating that it had once provided Rutherford with a piece of scrap paper on which to write an urgent memo, I opened it carefully, its crispness promising that I was probably the first to have set eyes on it. My suspicion turned out to be right. Inside, in Rutherford's bold hand, was written in pencil "Madam Curie says 5 gms [of radium] should be enough." Enough for what I wondered? No one will ever know, but that unexpected discovery was for me more precious than seeing his "official" notebooks.

The rays from radium today are best known as a treatment for cancer, but undirected can cause great damage and suffering. Marie Curie began to suffer from strange illnesses. Though she survived to the age of 67, her hands were wrapped to protect the blistering and she eventually died of aplastic anaemia, a condition produced by overexposure to radiation. Her experimental notebooks, and even her cookery books, were radioactive 50 years later.

The Curies shared the 1903 Nobel Prize with Becquerel. It was he who had discovered the phenomenon, but it was the Curies that had realized its awesome potential. It was Rutherford who would use it as a tool to reveal the nature of the atom.

In 1905, Rutherford began to investigate the radiation of alpha particles from thorium, radium, and other elements. He found that the alpha particles emitted by each of these elements had the same mass as that of a helium atom, but unlike normal helium atoms, the alpha particles carried two units of positive charge (in a convention where an electron carried one unit of negative). This led him to suspect that alpha particles are doubly ionized atoms of helium. This suspicion was reinforced by the fact that helium is found trapped within crystalline ores of uranium and thorium—a hitherto unexplained phenomenon but one which would be natural if the uranium or thorium were spontaneously emitting alpha particles, helium.

This was good circumstantial evidence that alpha particles are the seeds of helium atoms, but it was not yet a conclusive proof. That came in 1908 and, in typical Rutherford style, was disarmingly simple.

By that time Rutherford was in Manchester where in October 1907 he had taken up post as professor. He had the departmental glass-blower prepare a tube of thick glass containing within it a second tube of thinner glass. The critical thing was that the thin glass would allow alpha particles to pass through whereas the thicker surrounds would trap them. The apparatus was first evacuated and then the inner tube filled with radon gas, which is a powerful emitter of alpha particles. The outer tube gradually filled with alpha particles whose positive charges attracted electrons to them, neutralized them, and formed atoms of helium. We know that because after some days Rutherford was able to energize the gas until it glowed and then made a spectrum of its light. The spectral lines are like a fingerprint, uniquely identifying the culprit as helium. This confirming experiment was made in mid-1908 and Rutherford announced it in his Nobel acceptance speech.

It is doubly ironic that he should have done so. When he had started at Manchester he had drawn up a list of research questions for a focused attack on the nature of the atom and its radiations; alpha particles were high on the list. The proof that alphas are doubly ionized helium atoms was high priority, hence his pride at announcing it in Stockholm. Lower on the list was what appeared to be a rather mundane topic entitled "Scattering of alphas particles." Far from being mundane, this would turn into a discovery even greater than that for which he won his Nobel Prize.

The nuclear atom

Take a deep breath; your lungs go up and down but your hair stays in place. This trivial observation has a profound implication.

Electrons are present in every atom of your body, every atom in the atmosphere, and in every atom of oxygen that you inhale, (which then mixes with carbon from your body and is exhaled as carbon dioxide). Every five seconds, on average, you are breathing in and then exhaling over a million million million such atoms, and hence even more negatively charged electrons. This is a huge amount of electrical charge, and yet your hair manages not to stand on end. It is a demonstration that all atoms contain, in addition to their negatively charged electrons, an exact counterbalance of positive charge. The once simple indestructible atom has been revealed to be an object with a detailed inner structure.

Rutherford and Soddy's discovery that elements transmute from one to another was the first clue that atoms are made from smaller pieces that are common to all and that it is the rearrangement of these that caused the alchemy. Thomson had already hinted as much with his 1897 assertion that atoms contain negatively charged electrons and that these are a common feature of all atomic elements. This insight begs the question of how atoms are constructed: how are the positive and negative charges arranged? What causes transmutation? What is the secret power of radioactivity?

The revelation of the inner atom, hardly dreamed of in 1894, was by the turn of the century the emerging challenge. With brilliant directness Rutherford went straight to the heart of the problem.

To understand better the story and how he did it, let me give the punch-line first. We know today that atoms contain negatively charged flighty electrons on the outside, remote from a central, compact, bulky nucleus that is positively charged. The attraction of opposite charges holds the electrons in place and produces intense electric fields. The nucleus is so small and the electrons so remote that, apart from these electric fields, the atom is 99.9999999999999 per cent empty space! The logo of electrons encircling the central nucleus adorns atomic institutions worldwide and makes the atomic structure familiar to us all today. This lopsided asymmetric structure is the key to life and existence and will be central to our story. How did this truth emerge?

That the atom is not a hard permanent structure but is substantially empty space was the first piece of the puzzle to fit into place. Philipp Lenard, the same who had missed the discovery of X-rays in 1893 (p. 88), was by 1903 bombarding atoms with cathode rays. He found that the rays passed through almost as if nothing was in their way. What a paradox: solid matter to the touch is transparent on the atomic scale. Lenard summarized the enigma with the remark "the space occupied by a cubic metre of solid platinum [is] as empty as the space of stars beyond the earth." As with his earlier failure, once again Lenard has seen a great truth but dimly, and the full glory was to go to others.

Atoms may be mostly empty but something defines them, giving mass to things. That there is more than simply space became clear in 1906 as a result of Rutherford's experiments with alpha particles.

Ever since Rutherford had first isolated the alpha particle in 1899, he had been working to establish its identity. As a spin-off he discovered that alpha particles are good atomic probes.

The alpha particles were moving faster than a speeding bullet, yet a thin sheet of mica deflected them slightly. Rutherford calculated that the electric fields within the mica must be immensely powerful compared with anything then known. Fields of such a strength, in air, would give sparks and the only explanation that he could think of was that these powerful electric fields must exist only within exceedingly small regions, smaller even than an atom. From this came his inspired guess: these intense electric fields are what hold the electrons in their atomic prisons and are capable of deflecting the swift alpha particles. Rutherford's genius once more came to the fore as he realized that he could now use the alpha particles to explore inside atoms, from the manner of their deflections he would be able to deduce the atomic electrical structure. Thus the scattering of alpha particles became one of the research problems on Rutherford's 1907 list as he started at Manchester.

Scattering the alphas was one thing; seeing and recording them was another. He coated a glass plate with zinc sulphide which scintillates (emits flashes of light) when struck by alpha particles. These flashes are very faint and with a microscope it is just about possible to detect individual alpha strikes, but it is hard on the eyes and concentration. Rutherford and his assistant, Hans Geiger, took it in turns to count the flashes. They would take half an hour to adjust their eyes to the dark and then record the feeble flashes for one or two minutes at a time. In their research notebooks one can see the rough writing of Rutherford interlaced with the finer hand of his collaborator as first the one and then the other would take measurements.

They fired the alpha particles at aluminum, platinum, and gold and painstakingly recorded the results. Yes; there were indeed intense electric fields, the pattern of deflections confirmed that, but there was an incessant problem of stray scattered alphas that they could not explain. Rutherford wondered whether these were being deflected from the metal surfaces of the apparatus. This seemed most unlikely as it would require scattering through a large angle to get these stray alphas into the region of interest to the main experiment; nonetheless, the stray alphas were real and had to be checked. He told Ernest Marsden, a young assistant, to see if any alpha particles could be bounced back from thin foils rather than passing through them.

To do so Marsden used an alpha particle beam directed at 45 degrees to a thin foil of gold. Geiger had invented an electrical counter that would emit an audible click each time an alpha hit (a prototype of the modern Geiger counter). Marsden placed one of Geiger's counters at 45 degrees to the foil so as to detect any alphas that "bounced" through a right angle. Of course, he had to ensure also that none came straight from the source to the counter without hitting the target, and so he placed a lump of lead between the source and the counter. Alphas would not get through lead and so any that he recorded would have to have been scattered from the foil. (Recall that the intense electric fields in atoms had only deflected the alphas by a small amount, whereas Marsden was now looking for alphas deflected by 90 degrees.) He expected to see none and so was immediately surprised to find alphas being scattered into his detector. This showed that fast-moving alphas packing a considerable punch could nonetheless be turned through a right angle. He double checked, triple checked, until he was sure he had overlooked nothing and prepared to tell Rutherford. Later, Rutherford described his shock with his famous remark that "it was as if you fired a 15 inch shell at a piece of tissue paper and it came back and hit you." To be deflected through 90 degrees after hitting a gold foil that is only one ten thousandth of a centimetre thick, as Marsden's results showed, required that they had been subjected to electric forces that were a thousand million times stronger than anything then known and far beyond anything that Rutherford and Geiger had been measuring.

Rutherford then spent a year deciding what the source could be and what these results implied. Even after he had proposed his radical explanation—that the positive charge of an atom resides on a compact massive nucleus—he seemed strangely reluctant to advertise it.

Whereas the light electrons hardly disturb the alphas, the positive charges are on a massive ball at a point in the middle of the atom. For a heavy element such as gold this "nucleus" is much heavier even than alpha particles and resists the invaders by electrical repulsion. "Like charges repel" is the rule and the bulky gold nucleus stays fast and disturbs the incoming alphas. The nearer the alpha particle is to make a direct hit, the more it is deflected. In a head-on collision it stops momentarily and then comes right back.

In reality the nucleus is very small and most of the alpha particles go straight on or nearly so; it's very rare that they bounce out at a large angle. These rare near hits were the source of the large deflections that Marsden had measured and which had been revealed initially as the stray scatters in Rutherford and Geiger's first experiment.

Rutherford calculated that the nucleus contains almost all of the atom's mass but that it occupies less than a thousandth of a

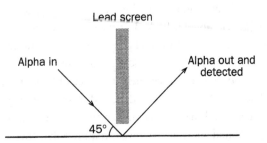

Fig. 7.1 Alpha particle scattering: the alpha particle comes in from the left and bounces off to the right where it is detected. A lead plate in the middle prevents a direct path from the alpha source to the detector.

millionth of a millionth of its volume. His astonishment was so great that in his notes (Fig. 7.2), preserved in the library at Cambridge University, his handwriting became almost totally illegible as he wrote "it is seen that the charged centre is very small compared with the radius of the atom." As remarked earlier, the atom is 99.99999999999999 per cent empty space, which is astonishing not least because solid matter seems so impenetrable.

In this we have essentially the picture of the nuclear atom that has survived for the subsequent 80 years. The neutral atom balances negative electrons and positive nuclei; in mass it is no contest as the positive nuclei outweigh the negative electrons by many several thousands. Our matter is composed of "Brobdignagian" positives and "Lilliputian" negatives; negative and positive electrical charges balance neatly but in a very lopsided asymmetrical fashion.

This inbuilt asymmetry at the heart of matter is crucial to existence as we shall see in subsequent chapters. First though, let's spend some time with the nucleus that Rutherford has discovered.

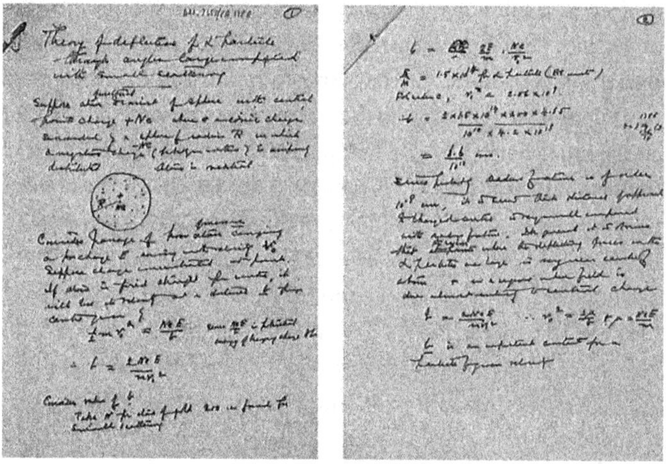

Fig. 7.2 Rutherford's notes where he calculates the size of the atomic nucleus.

It has a rich structure of its own and clever exploitation of its properties has helped forensic detectives in their quest for the secret of life's asymmetries.

The discovery of the proton and neutron

Rutherford knew that the nuclei of all atomic elements contained positive charge and that heavy elements, such as iron, lead, and gold, have more than their lighter counterparts such as hydrogen, carbon, and nitrogen. Moreover, the amount of charge increased by unit amounts as one moved from one element to the next in the periodic table, starting from hydrogen and working up towards uranium. It seemed that there might be some positively charged entity that was common to the nuclei of all atoms. This we now know to be true; it is a particle that Rutherford discovered and named the "proton" (from the Greek for "first").

Rutherford had been doing the work leading to this discovery during the years of the First World War, during which time he had been in great demand by the British Admiralty. He had played a seminal role in applying science to the war effort, devising ways of detecting submarines (among a host of other problems). That he realized the singular importance of his emerging discovery on the structure of atomic nuclei is shown by the fact that he excused himself from a meeting at the Admiralty with the excuse that "if he had achieved what he suspected, then this would be more important than the war"—prophetic indeed and, in view of what would occur 25 years later in terminating the Second World War, ironic.

Until 1932, the proton was the only established component of the atomic nucleus. The discovery of the neutron, a particle with nearly the same mass as the proton, changed that, clarified the mystery of the periodic table and isotopes, and provided a shattering element into nuclear science. That neutron has proved to be a singular tool. Having no electrical charge it is impervious to the electrical barrier surrounding atomic nuclei and can pass through and enter into the nucleus. The neutron, perhaps more than any other entity, has opened the nucleus up for examination.

As a student in 1911, James Chadwick had been present at the Manchester Philosophical Society meeting on 7 March at which Rutherford had announced his discovery of the atomic nucleus.

The idea of the neutron, a neutral particle similar to the proton in all but its electrical charge and the third major constituent of atoms besides electrons and protons, was first articulated by Rutherford. It was at the Bakerian Lecture of the Royal Society on 3 June 1920 that he speculated on the possibility of "an atom of mass 1 which has zero electrical charge." He did not envisage it as an elementary particle but as a composite—an electron and proton tightly united. He noted that it would be able to move freely through matter and, not being subject to the electrical repulsion that alpha particles or protons are on encountering the atomic nucleus, could be the most effective tool to penetrate and probe the nucleus.

By this time Chadwick had become Rutherford's assistant. He attended the Royal Society lecture and heard Rutherford propose the existence of the neutron, the missing piece in the nucleus' construction—an experience that proved seminal for Chadwick as 12 years later it would be he that won the prize of proving Rutherford right by finding it. By contrast, Irene Joliot-Curie (daughter of Marie and Pierre Curie) did not read the transcript of Rutherford's speculation (as it is unusual for the Bakerian Lecture to contain new ideas) and this chance may have caused her to miss the discovery that would be hers for the taking in 1931.

When Rutherford proposed the neutron, less than a year had passed since he had discovered the proton. There was however still a nagging possibility that contamination was deceiving them, and so Rutherford guided Chadwick into helping him to tighten up the experimental evidence. Chadwick had been developing a microscope that would improve the tedious counting of the crucial scintillations and, by 1920, they were ready to start taking measurements together. To prepare their eyes for the task of seeing faint flashes, they would sit together in a dark room for half an hour before starting their work. During this time the pair talked at length; Chadwick was tutored in detail on Rutherford's intuition about the neutron and the reasons why he insisted that it must exist.

Rutherford was concerned about how each element was constructed. Painstaking chemistry has established the periodic table of the elements; starting at hydrogen, then helium, and so on, each element had its place on the series known as the "atomic number." Rutherford's investigation of atomic nuclei had shown that the atomic number was the same as the amount of positive electrical charge on that element's atomic nucleus relative to that of the proton. Hence the atomic number corresponded to the number of protons in the nucleus. So far so good. However, the relative weights of the elemental nuclei had also been measured and, as one moved up the periodic table, the magnitude of these weights grew faster than the atomic number, being almost double by the time one reached iron and nearly triple for uranium. For example, helium has atomic number 2 and weight 4, nitrogen has atomic number 7 and weight 14, and as one moved along the periodic table the difference grew until by uranium with atomic number 92, sometimes one found weight 235 but more often one found 238. Rutherford suspected that electrically neutral twins of the proton also were present in the nucleus; the more of them, so the greater the mass without changing the total charge. This would explain the discrepancy between atomic number and atomic mass as being due to the number of neutrons. The idea could also explain the phenomenon of isotopes where a single element, such as uranium, could occur with more than one atomic weight: the number of neutrons in the two cases must differ. From the darkroom conversations with Rutherford, Chadwick became convinced that the neutron must exist. The problem was how to prove it.

The critical step on the route to eventual discovery came from Walther Bothe and Herbert Becke in Germany in 1928. They were bombarding light elements with alpha particles. In the course of this they bombarded beryllium, a light element, and found that, apparently, gamma rays were emitted more intensely than anything they had seen before. What was more strange was that the energy of the radiation appeared to be greater than the laws of physics would allow, unless the beryllium nucleus was disintegrating.

The results were startling and two groups set to work to replicate the phenomenon. In France, Irene Curie collaborated with her husband, Frederic Joliot; meanwhile in England, Chadwick assigned the task to a student.

A critical discovery ensued in mid-1931 when the student noticed that the radiation from beryllium was more intense in the forward direction (that is, the direction of the incident alphas) than to the rear. This is what one would expect if the radiation consisted of massive particles that resided initially within the beryllium and were ejected by the incident alphas. By contrast, gamma rays have no mass and like light emitted from a lamp should be radiated in all directions equally. Chadwick was convinced that this was the neutron and that the Germans had missed it.

Meanwhile Joliot and Curie in France were also replicating the German work. Their advantage consisted of having access to large quantities of polonium, a powerful source of alpha rays; Marie Curie's Radium Institute received spent radon capsules that had been used in cancer treatment. Once the radon had decayed it was no longer useful to the doctors but its decay products included polonium 210. The Joliot-Curies had developed chemical techniques for purifying polonium and in Paris had accumulated more of it than anywhere else in the world.

The polonium created a beam of alphas which in turn blasted beryllium and ejected the mystery radiation. Irene Curie announced to the French Academy of Sciences on 28 December 1931 that she too had seen this radiation and that in her experiment it was even more intense than the Germans had found. Now Joliot and Curie took a new step. They placed other materials downstream of the beryllium to see if the mystery rays would knock protons out of matter as alpha particles did. They were successful here too; the mystery radiation rejected protons from within the material in a way similar to a cue ball knocking a snooker ball out of the pack. The critical thing for collisions in snooker is that balls have the same mass; the response of the proton to the mystery radiation was as one would expect if the radiation consisted of particles whose mass is similar to that of the proton. In retrospect, it is astonishing that Joliot and Curie failed to realize that the mystery radiation was not gamma rays but massive neutrons, and so also missed the Nobel Prize that would have followed. One reason for their failure was that they had not read Rutherford's Bakerian Lecture and so were unaware of the idea that there might be a massive neutral particle. Consequently, knowing only of gamma rays, like the Germans before them, they misinterpreted what they were seeing.

It was early in February 1932 when Chadwick received news of their work and he immediately realized that Joliot and Curie had made a fatal error. Gamma rays have no mass and although they may easily knock out lightweight electrons from atoms, they would hardly disturb a proton that is nearly 2,000 times heavier. He told Rutherford of the French claim and Rutherford exclaimed that although he believed the observations, the interpretation must be wrong.

Convinced now that the neutron was there to be found, but needing to prove it and to do so before the French team realized their error, Chadwick set to work feverishly. Starting on 7 February he worked day and night for 10 days before he had proof; it was one of the most rapid, intense, and far-reaching discoveries in physics.

Chadwick had a radiation source of alpha bullets and duly set out first to reproduce what Joliot and Curie had done in order to check that his experiment worked. Having done so he then made the critical breakthrough: he placed a whole range of elements in the path of the mystery radiation to see what would happen. The mystery radiation ejected protons from each and every element that he tried, even from the gold in Rutherford's Nobel medal. In each case a similar number of protons were ejected and, crucially, their energies were larger than could be possible had the mystery radiation been gamma rays.

The basic point, as Chadwick realized, was that if the culprit had been gamma rays hitting heavier and heavier targets, it would have needed ever higher energies to disturb them and eject the protons. However, if the mystery radiation consisted of neutral particles of mass similar to that of a proton, then like the collision of two snooker balls, the protons would recoil the same, independent of the size of their original elemental nuclear home.

Thus, by means of beams of naturally occurring alpha particles, the neutron had been prised loose. In a mere 10 days and nights Chadwick had established the third basic constituent of matter. Its existence immediately explained why atomic weight differs from atomic number. The latter, the position in the periodic table, is determined by the nuclear charge or number of protons. The neutron added mass to the nucleus without addition of charge.

The discovery of the neutron was so important that it has been widely regarded as the defining moment when the science of nuclear physics truly began and an understanding of the structure of matter emerged.

The atomic nucleus

The picture is this—the atomic nucleus consists of protons and neutrons; the protons have positive electrical charge that attracts electrons remotely to build atoms; the neutrons add mass to the nucleus but do not alter its electrical charge and so do not influence the number of electrons that are gripped remotely; the neutrons help to stabilize the nucleus.

Atoms of a particular element can have different numbers of neutrons in their nuclei; these forms of the same element are called isotopes. On earth the relative abundance of the various isotopes of the elements is well known; elsewhere in space, the abundance can differ. The abundance of the isotopes is like a fingerprint that can identify the origin of the material—whether it is native to the earth or came from outer space for instance. It was by such means that the extraterrestrial origins of the Murchison meteorite (page 76) were established.

It is the positively charged protons that grip the electrons making atoms. The atomic structure is balanced in electric charge, the positive of each proton and the negative of each electron being in perfect harmony, but it is lopsided in mass. Each proton or neutron is some 2000 times heavier than an electron. The result is that the nucleus is bulky and static whereas electrons are light and flighty. It is these electrons that move from atom to atom, building up molecules and determining their shapes. We have seen how these shapes can form "isomers" (mirror images) and how life chooses certain isomers while rejecting others. Later on we will meet ideas on why it has turned out like this, why our DNA and amino acids coil in one sense whereas their mirror images are not seen, at least in living organisms.

What has only recently been realized is that all of these asymmetric structures have probably emerged from an original state of symmetry. The structure of the atom, painstakingly established at the start of the twentieth century, is how matter is when in the relatively cold conditions that we are used to. In the heat of the stars or the extremes of the Big Bang, things were very different.

There is a similar story for the forces of nature that glue the fundamental pieces together; it turns out that these, and even the basic particles themselves, have evolved from a unity to a variety of different forms today. The rich structures that are familiar today have emerged from a symmetry in the early moments of the universe, the singular occurrence that we call the Big Bang. As we go back in time towards that moment, we begin to glimpse that symmetry and perhaps the true harmony of the universe. That is the journey that we will start in the next chapter.

Chapter 8

A glimpse of symmetry

Guided by the stars in the hour before dawn, fishermen catch the tide to bring them home; amateur astronomers gather at a prearranged hour in a remote part of the world to see day turn to night as the moon eclipses the sun; satellites orbit above us; spacecrafts take humans to the moon or use Jupiter's gravity like a slingshot to explore the solar system. All of these, and much more, depend on the force of gravity. Ever since Isaac Newton realized that falling apples, the tides, and our relations to the stars are all governed by the same universal force, we have known the rules and been able to calculate the timing of astronomical events with certainty.

If, like me, you suffer vertigo as you look over the edge of a high building, you will not need convincing that gravity has a powerful pull. At an early age we know that it is inadvisable to step off the edge of cliffs, though it is somehow miraculous that the earth "knows" that we are there, remote and momentarily suspended in mid-air before being accelerated into a final nosedive. Even more remarkable is that the sun, no more than a thumbnail in size as viewed from earth, can entrap us and the planets in a cosmic waltz around the vastness of space hundreds of millions of kilometres distant.

Newton's seminal insight was that gravity's pull between two bodies diminishes as the square of the distance between them increases. The fact that it behaves like this is crucial. We are trapped on the earth that orbits the sun and the gravity of distant stars is trifling, but if it had been the case instead that the force of gravity was independent of distance, the tug of immense galaxies at the far reaches of space would have ruled. As it is, individual remote galaxies don't affect us; the timings of the tides, eclipses, and of voyaging spacecraft are governed by the gravitational influences of local planets and moons and can be accurately determined. The reason that the force of gravity diminishes with the square of the distance is intimately due to the three-dimensional nature of space and the fact that gravity fills all of it. A massive body, such as the earth or the sun, somehow sends out its gravitational tentacles into space, in all directions, uniformly.

The earth's orbit is very nearly circular. Imagine the sun at the centre of a ball whose diameter is the same as that of the earth's orbit. The gravitational tug on our planet is the same at all points during this annual orbit and, indeed, the same as at all points on the surface of the imaginary ball. If we now imagined ourselves transported to an orbit that was double that of the earth's, the surface of the imaginary ball would be four times greater (the area increases with the square of the distance). Newton realised that if the force of gravity were likened to tentacles spreading out from the source in all directions symmetrically, then the intensity at any distance would be spread uniformly across the area of the imaginary ball. As the area increases with the radical distance squared, so will the intensity at any point on it correspondingly weaken.

This analogy highlights the intimate relation between the behaviour of the gravitational force and the three-dimensional nature of space, which has been known since Newton. It gives us an important clue to the mystery of how a force can occur between apparently disconnected bodies: the intervening space is somehow directly involved. This idea also comes from Newton and its essential features have remained with us for 300 years, enriched by the insights of Einstein and applied in ways that Newton never knew. The basic idea is that there is a kind of tension existing in otherwise "empty" space that manifests itself by producing forces on objects that happen to be in the vicinity. This tension's sphere of influence is called a field and it is the earth's gravitational field stretching into space that pulls skydivers to ground.

Likewise, compass needles swing in the magnetic field of the earth. The molten metal core of our planet swirls as we rotate, the heat disrupting its atoms such that their electrons flow freely: the molten core is alive with electric current. These currents make the earth a huge magnet with north and south poles, and magnetic arms that stretch out into space. Far stronger than the earth's gravity, its magnetic field will deflect a small compass needle almost instantaneously as we move it around. This phenomenon has been the guide for travellers since time immemorial and is another mysterious example of apparent action at a distance. How does a little magnet, perhaps no larger than a centimetre, "know" which way is the north pole, perhaps several thousand kilometres distant? As with gravity, we are used to the idea of a magnetic field stretching throughout space, its intensity dying away at a great distance.

Electric and magnetic forces do more than just make magnetic fields for the use of compasses; they are intrinsic to everything that we see and fundamental to our existence. We ourselves are held together by these electromagnetic forces among our constituent atoms and molecules—the force between two hydrogen atoms, for example, being some 10^{40} times more powerful than their mutual gravitational attraction. The gravitational attraction between your head and your feet is very small; it is the electromagnetic forces that give your limbs their form and structure.

The rule that we learn at school about electric charges is that opposites attract and like charges repel. There are both types within typical atoms, the negatively charged electrons being on the periphery, while at the centre we have the positive nuclear core. When atoms are close to one another, the positively charged nucleus of one atom can attract the negatively charged electrons of a neighbour, causing the two atoms to move a little closer to one another. As a result, groups of atoms can be mutually ensnared and clump together forming molecules. Chemistry and life are thus choreographed by the electromagnetic force.

The individual rocks of the solid earth are glued by electromagnetic forces, giving shape and form to the mountains and canyons on its surface. Although the gravitational attraction between individual atoms is nugatory, the contribution from each and every atom in a large object adds up. Gravity rules once an object is larger than about 500 kilometres in diameter. Collectively, it is their weight, the force of their gravity, that holds the spherical whole together. Gravity ensures that large objects are spherical; small objects are not. As we have already noted in Chapter 2, were mountains to be forced upwards too far, their weight would break the electromagnetic bonds between their atoms: the rocks would melt and the mountains would sink. This is why there are no mountains as high as 100 kilometres and the earth overall, viewed from afar, appears smooth and round. Smaller structures, by contrast, have their shapes determined by the way that their component atoms are joined and the resulting patterns are richer than the bland sphere.

The forces of nature

The fundamental forces of gravity and electromagnetism are the two that are most familiar in our day-to-day existence. However, in addition there are two other forces that act in and around the atomic nucleus. One of these is stronger than the electromagnetic force; the other is weaker. As a result they have become known as the strong and the weak forces though, as we shall see, this can be something of a misnomer. This quarter of forces controls our life.

Picture ancient Druids at Stonehenge as the sun rises; all four forces are at work. The first, *gravity*, sticks the stones on the surface. The dawn light that shines on them, as night turns to day, is a result of the *electromagnetic* force. The sun's surface brightness is the remote product of thermonuclear reactions taking place at its centre: its hydrogen core is turning into helium as a result of protons converting into neutrons in various ways through the *weak* interaction. (This is necessary to generate the energy that is eventually transmitted across space as sunlight and which provides just enough warmth here that the chemical processes essential to life can take place.) That there is any carbon, or other elements of life, is due also to the fourth force—the *strong* force—without which there would be nothing but hydrogen in the universe.

Hydrogen is the simplest element, the commonest in the universe at large, and the stuff of the stars. Hydrogen is very rare on earth, by contrast, floating to the top of the atmosphere and out into space once released from its usual prison, the molecule of water, H_2O. The earth, our home, is rich in the other elements including carbon in our bodies, nitrogen and oxygen in the air, and gold in the bank! These elements are all made of atoms at the heart of each being its essential seed, the atomic nucleus, containing a large number of tightly packed protons. The golden rule of electrical forces is that "like charges repel" and so it is at first sight a paradox that so many positively charged protons can stick together. That they do is because of the strong force, a powerful attractor between neutrons and protons, once they touch, that overrides the electrical disruption.

All four of the forces are needed for life to emerge. To recap: the strong force compacts atomic nuclei; the weak helps transmute the elements within stars to build up the richness of the periodic table; the electromagnetic ensnares electrons and builds atoms and molecules; while gravity enabled the star to exist in the first place. It is remarkable that just these four forces are needed for us to be here. The strengths and characters of these forces are tightly related. For example, had the weak force been just three times more powerful than in reality, the life-giving warmth of the sun would have ended long ago; alternatively, had the force been more feeble, it is probable that the elements of life would not yet have been "cooked" within the stars.

We are beginning to understand the interrelationships among these forces but not why their relative strengths are as they are—though as we shall see later, we have some clues. There are various symmetries among them that we are starting to discern.

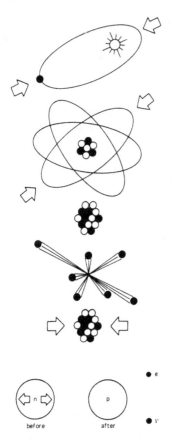

Gravitational force
This acts between all particles. Pulls matter together. Binding force of the solar system and galaxies

Electromagnetic force
Unlike charges attract. The cloud of negatively charged electrons is held around a positively charged nucleus. Binding force of atoms.

Strong force
The nucleus of an atom contains protons. Like charges repel by the electromagnetic force. But the nucleus doesn't blow apart because the strong force attract protons and neutrons and binds the nucleus.
    This is now believed to be a remnant of a more powerful colour force acting on quarks inside the protons and neutrons.

Weak force
This causes radioactive decay of some nuclei. A neutron breaks up, emits an electron and becomes a proton. When operating in this way this force is a thousands times weaker than the strong nuclear force. At high energies it is not so weak and begins to act in a similar way to the electromagnetic force. At very high frequencies the electromagnetic and weak forces appear to be intimately related.

Fig. 8.1 The four fundamental forces.

Inside the sun

The belief that there is unity among widely differing phenomena has been a successful drive behind science through the ages. That falling objects, the tides, planetary orbits, and the motions of entire galaxies are all manifestations of universal gravity, was one of the first great insights. Comb your hair rapidly on a dry day and sparks may jump; suspend a magnet from a thin thread and it will swing to the north–south axis: it is not at all obvious that these electric and magnetic effects are related, yet Maxwell (page 83) realized that they are and created his unified description of electromagnetism.

Today, electromagnetism encompasses the structures of atoms and molecules and the fields of electronics, optics, and telecommunications. When Becquerel, the Curies, and Rutherford first recognized beta radioactivity, no one imagined its relation to a rainbow—but we now know about it. We will come to that later; the symbiosis between the weak force of radioactivity and the electromagnetic force is only manifest under conditions of energy or temperature far beyond those we are used to day-to-day. Following this discovery in the 1980s, physicists searching for the "holy grail" of the ultimate theory have increasingly focused their attention on the extreme heat of the early universe in the belief that it is there that the perfect symmetry in the natural laws is to be found.

If there is a credo of the Creation in modern cosmology it is that the universe was initially symmetric; had it appeared otherwise, our aesthetic sense is to wonder "what ordered that?" and to seek a deeper explanation. This belief in an initial symmetry has grown as we have recognized ever more examples suggesting that asymmetries in the universe "now," such as the radically different strengths and discriminatory properties of the fundamental forces needed for life, are descended from more symmetric conditions earlier in the history of the universe. We know that initially the universe was hotter than any star and is now and immeasurably hot compared to the ambient temperature of deep space "today" (a chilling $-270°C$—a mere $3°C$ above absolute zero).

In conditions of such extreme heat, matter behaves in unfamiliar ways. For example, the sun is quite unlike the earth. In our world protons and neutrons grip together in tight nuclei whose electrical charges ensnare passing electrons to form atoms. These atoms are the foundation of bulk matter—in solid rocks, liquid oceans, and gaseous atmosphere, among the myriad of assorted compounds that comprise the beauty of our planet. In solid, liquid, and gas the atoms are increasingly free, but they are the same atoms; whether steam, rain, or snow the essence is always H_2O. In radioactivity, the nuclear seeds alter as the atomic elements transmute, but electrons are ensnared such that atoms are still recognizable. As temperatures rise, however, the atoms move faster and collide with increasing violence. Above a few thousand degrees, as in the sun, the violence is enough to knock all of the electrons completely out of their atoms. At such temperatures atoms cannot survive, their individual constituent particles, electrons, and protons in hydrogen, for example, flow independently as two electrically charged gases. This state of matter is called "plasma."

The universe now has innumerable hot pockets like the sun, with much larger cool regions in interstellar space; life could not occur in either of these extremes. Life needs some warmth to encourage an efficient metabolism, growth, and replication of molecular DNA, but not too much or the atoms would be torn asunder into their component pieces, as in the stars. It has taken time for the universe to provide the right conditions. Today we inhabit a universe that has evolved for some 15 thousand million years since its birth in the extreme heat that we call the "Big Bang." During the first moments, the universe was initially hotter than our sun is now. Looking deep into space with a telescope, and back in time (for it takes light time to travel across space from the stars to our eyes), we can peer to within 100,000 years of the Creation. However, we cannot see into the Big Bang itself because the heat is so intense that it would be like trying to look into the sun: as our eyes cannot see beneath the solar surface, neither can they peer into the hot veil of the early universe. However, all is not lost, as we can piece together what the conditions are (were) like by noting the tell-tale messages that escape, and then make forensic examination of these clues here on earth.

We can deduce what the inner recesses of the sun are like by detecting neutrinos—ghostly, electrically neutral particles that are produced in the nuclear furnace at its centre and fly out unimpeded, pouring through space at or near the speed of light. On reaching the earth they can be captured in specially designed detectors, their numbers and range of energies giving information about the conditions where they were born. Nuclear physicists can replay, under controlled conditions, the reactions among protons and other atomic nuclei that occur at the ambient temperatures of the sun. The propensity for nuclei to meet, transform, release, or absorb energy are all measured in laboratory experiments here and the information used in comparing with the models that astrophysicists have for the interiors of the sun and other stars. It is as if the stars are sending us fax messages by their light and sub-atomic particles and it is the intelligence gained from laboratory experiments here on earth that enables us to decode them.

The higher the temperature, the more violent are the collisions among the particles in the plasma. The impact can tear nuclei apart, releasing energy that was locked in since their first creation. For example, deuterium and tritium are heavier forms, isotopes, of hydrogen, whose nuclei contain a proton (as does hydrogen) but bound tightly to one or two neutrons respectively. At Culham in England, a European team is attempting to create a self-sustaining fusion reaction on earth more efficient than in the sun. To do so they collide nuclei of deuterium and tritium at energies corresponding to temperatures of some 100 million degrees—a factor 10 times greater than in the centre of the sun. Under these conditions the collisions disrupt the nuclei, releasing energy which can then be used to help power the experiment. The challenges are in the engineering and in controlling the wild gyrating electrical plasma, whereas the fundamental behaviour of the nuclear collisions at these temperatures is by now a familiar and well-understood science. The nuclei break up into pieces, protons and neutrons rearranging themselves to form more stable combinations.

The sun is going through the simplest example of this as protons, its essential substance, collide and fuse together, aided by the weak interaction that converts protons into neutrons (emitting the ghostly neutrinos as one of the by-products), eventually building up the nucleus of the next heaviest element, helium. The sun is burning up protons at the rate of some 600 million tonnes each second, creating helium as the ash; four billion tonnes of matter are converted into energy which is released as sunlight. Some stars, having exhausted their primeval proton fuel, live on by helium nuclei merging to build up yet heavier elements. By colliding the nuclei of helium and other elements at laboratories on earth and at the energies (temperatures) of the stars we have pieced together the way that the stars are born, live, and die. They are "cookers," far more advanced than the microwave, the ingredients originating as hydrogen which is baked into heavier elements including those essential for life. As a result of these insights we can now successfully explain the relative abundances of the elements; thus the ubiquitous prevalence of carbon and oxygen is understood as is the scarcity, and hence value, of gold.

There are several details to be solved, such as what goes on in the rapid cataclysmic explosion of a supernova, where very short-lived radioactive nuclei may fuse, fragment, and reform before being ejected in a final stable form into the cosmos, perhaps to fuel planets and stellar systems like ours. To know more of this will require experiments where rare and unstable nuclear isotopes are brought into contact at the energies appropriate to the supernova explosion. The technical challenge is one of creating intense beams of these rare entities and this project is high on the agenda for nuclear physicists. However, the underlying rules are already known: at these energies the basic processes are the rearrangement of neutrons and protons like the partners at a barn dance.

Extraterrestrials

As the energy of the collisions increases, new things begin to occur. We know this from experiments on earth and are aware that it happens in the universe too as cosmic rays—little particles smaller than atoms—are showering down from the heavens as you read this. At this very moment millions of them are pouring through you every second without you being aware of them. They are the result of stars that exploded long ago, deep in space, with such force that they were utterly disrupted, ejecting some trillion trillion trillion trillion elemental nuclear particles into the galaxy. Electric and magnetic fields that permeate interstellar space whip the debris into violent motion, giving to some of the atomic particles energies more extreme than anything found on earth. A few, passing near our planet, are trapped by the magnetic arms of the earth and pulled down until they crash into the atmosphere far above our heads. The violence of these collisions breaks the atoms of the air into little pieces that shower down to the ground.

These cosmic particles are smaller than atoms but even so we can detect their presence. If you live far enough north you may have seen their effects with your own eyes. The Northern Lights or "Aurora Borealis" are the result of the cosmic rays having been concentrated into quite intense beams by the earth's magnetic poles, and shaking light from the atoms in the upper atmosphere. It is also possible to see some of the cosmic rays directly with relatively simple instruments. The particles can short circuit a chamber full of electrically charged wires; the ensuing sparks in this "spark chamber" reveal their flight paths. Very occasionally, if you are unlucky, they may confuse your computer chips or crash your system.

Cosmic rays have such high energies that their collisions with atoms in the upper atmosphere can create short-lived exotic particles whose existence had not been previously suspected let alone known to scientists trapped in their earthbound laboratories. Any theory of the universe must explain why they occur. It was the discovery of such exotic, "strange" particles half a century ago that stimulated the modern science of high-energy particle physics, where beams of protons or electrons are accelerated to high energies, simulating the cosmic rays.

These rays have brought us news of a violent universe, but they arrived at random and raised more questions than they answered.

Who ordered these particles? Why are they as they are? How are they related to the familiar proton and neutron that are the seeds of life as we know it? Are they necessary to the existence of the material universe? And so on. To answer these and a host of other questions the era of "atom smashes"—huge accelerators of particles capable of reproducing the energies of the cosmic rays and creating the fleeting exotic particles under controlled conditions—began. This quest originated with the intention of understanding the nature and structure of matter, but in recent years has evolved into a study of the early universe, where the temperature was enormous and the energy of the swarming particles correspondingly large. Modern accelerators, such as at CERN in Geneva, Switzerland, can regularly bring particles into collision at energies more extreme even than those at the core of the most exotic stars, but akin to those of the early universe. By such experiments with high-energy particle accelerators we are able to look ever deeper at the extreme conditions of the aftershocks of the Big Bang.

This is where particle physics (concerned with experimentation) and cosmology (theories on the structure and evolution of the universe at large) combine to give a more profound insight than either alone could do. Particle physicists look deep into matter and have identified the nature of the short-lived exotic particles in the cosmic rays. They have deduced why these and their siblings form such a rich variety: it is because the proton, neutron, and their strange relatives are not the fundamental pieces of matter. It is as if, in the first half of the twentieth century, we thought that we had found the basic letters of nature's alphabet only to discover that the multitude are constructable from combinations of dots and dashes as in Morse code. It was in 1964 that the Americans, Murray Gell-Mann and George Zweig independently realized that they could make such a code to describe all of the particles that experience the strong interaction. The analogues of the dots and dashes are small particles that Gell-Mann called "quarks" and Zweig called "aces"; the name quarks has stuck. This turned out to be more than a mere encoding; experiments probing deep inside the proton have seen the quarks within. Today we recognize that what we call "proton" is actually a swarm of quarks "glued" together by the strong force.

Like anthropologists, the particle physicists have studied the quarks and learned how they behave. It is in doing so that the first glimpse of a deeper symmetry has been discerned.

The strong force between protons and neutrons in atomic nuclei today is but a feeble remnant of the powerful forces that glue the quarks to one another. It turns out that in the first billionths of a second after the Big Bang it was the quarks that were the primeval seeds of the forms of matter that we now know; experiments have shown that at energies similar to those present at that epoch, the character of the strong force was different. Under those conditions the quarks roamed free; the glue that ensnared them following similar rules and *modus operandi* as the more familiar electromagnetic force. These are hints that two of the natural forces, the electromagnetic and the strong, follow the same rules, or did so when the universe was young.

Although these are only hints, at present, we are confident that we have stumbled on an underlying unity because we have caught nature at it elsewhere. We know that the electromagnetic and weak (radioactivity) forces are really united when the temperature is 400 billion times greater than body heat because by the 1990s we reproduced such conditions in the LEP accelerator at CERN (Chapter 11). This showed that the weak and electromagnetic forces, as we know them at room temperature, are merged as one at high energies; this united force is called the "electroweak" force.

To understand what this means and to begin to get a sense of how this, and other symmetries, have become hidden as the universe evolved, we need to see how forces work. I will now describe the evolution of ideas on how it is that the compass needle "knows" of the remote magnetic poles of the planet.

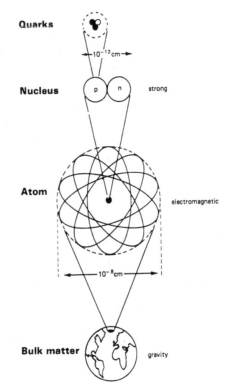

Fig. 8.2 Particles and forces.

Unified forces

One-third of a millenium has elapsed since Isaac Newton first
realized the paradox of light. On the one hand he noticed that it
behaved as if made of particles—photons—traveling in straight
lines, leaving shadows when interrupted, its brightness greater
when more photons were present. However, light also on occasions behaves in ways more reminiscent of waves that the photon
picture could not easily explain.

The wave picture of light came into major prominence again
when, in the mid-nineteenth century, James Clerk Maxwell

showed that light consisted of oscillating electric and magnetic fields: an electromagnetic wave. Light seemed to have a schizophrenic nature, sometimes behaving as particles, at other times as a wave. It was less than a hundred years ago when the paradox was resolved by the quantum theory, that has helped also to give a physical picture to the mysterious concept of the forces acting through a field. According to quantum theory, the field is not a passive thing but ripples across the intervening space; it is these ripples that move things about, transmitting the force. In the case of the electromagnetic field it is an electromagnetic wave that does the job. Quantum theory says also that these ripples are organized in little bundles that act themselves like particles—photons in the case of electromagnetic waves. It is these photons flitting between one electrically charged particle and another, bumping into them, being absorbed, emitted, reabsorbed, and re-emitted that cause things to be moved around and to respond to the "electromagnetic force." Action at a distance is transmitted by an agent—the photon. Similarly in the case of the gravitational force; the agent is known as the graviton.

The energy of each photon is governed solely by the wavelength of the lightwave: small wavelengths as in blue, ultraviolet, or X-rays have larger energy, while long waves, as in the red, infra-red, and radio waves, have relatively low energies. As the length of the wave tends to infinite extent, the amount of energy transported tends to vanish. In the most extreme case one can imagine a wavelength spanning across space with no energy in its photons at all.

The concept that an electromagnetic wave consists of quantum bundles, photons, underpins much of modern technology; the idea that photons transmit the electromagnetic force has been established by a range of precision experiments during the last 50 years. The strong force that glues the nuclear constituents together is transmitted by analogues of the photon, known as "gluons." (This lack of imagination in names is even more marked for the weak force where the quantum bundle of weak radiation, the agent responsible for beta radioactivity, is known simply as "W" for weak.) The W carries electric charge, either positive or negative; the theory that unites the weak and electromagnetic interactions successfully predicted that there would also be an electrically neutral analogue of the W. This zero charged object is called Z^0 (the ultimate in unimaginative naming—the Z is for zero and the superscript "o" tautologically reiterates this).

Where Newton had unified apples and planets with gravity, and Maxwell had united electricity and magnetism into electromagnetism, it was in 1979 that the Nobel Prize was awarded to Sheldon Glashow, Steven Weinberg, and Abdus Salam for their theory unifying the electromagnetic and weak forces into what is today known as the "electroweak" force. The price that they had to pay was that there should exist three varieties of weak radiation carried by quantum bundles (W^+, W^-, and Z^0) with the W and Z being very massive analogues of the (massless) photon. As a result, the Z^0 became whimsically known as "heavy light." In the 1980s experiments at CERN liberated the W and Z as real entities, proving Glashow, Salam and Weinberg to be correct. The Nobel committee awarded their 1984 physics prize to Carlo Rubbia and Simon van der Meer for their role in creating the CERN machine that produced the W and Z, and perhaps also with some relief that their earlier award to the trio of theorists, now proven to be correct, did not need to be revoked! The CERN experiments, one led by Rubbia and the other by Pierre Darriulat and Luigi Di Lella showed that the W and Z weighed in around 90 times heavier than a hydrogen atom.

The W and Z are much like the photon but for their huge masses. Had the W and Z been massless like a photon, the electroweak force would have been symmetric in all its facets; the strengths of the electromagnetic force and the force that triggers radioactivity would have been similar. The huge mass of the W and Z in contrast to the massless photon is what breaks the symmetry of the electroweak force, and makes the "weak" part of it truly feeble by comparison to the electromagnetic force. These bulky beasts are lethargic at room temperature, and barely active—hence the weakness of the ensuing force; by contrast, at the high temperatures of the early universe, which are reproduced at CERN, the particles are so agitated that heavy and light behave alike. The W and Z and photon become equally flighty, and the unity of the weak and electromagnetic force carriers becomes manifest.

Why it is that W and Z are so massive, whereas the photon has no mass at all, is one of the frontier questions of our time and will be the subject of later chapters, but it is worth reflecting for a moment on the fortune that it is like this. Had the weak force that can convert protons to neutrons in the sun not been feeble then, as we noted on page 143, the sun would have burned much faster and used up all of its fuel before the complex molecules of life had had time to form. Arcane though it sounds, it is nonetheless true that we and life on earth, five billion years after the first fusion took place in our sun, is because the W is so massive while the photon has no mass. So here again we find that our existence is critically dependent on nature having broken symmetry as the universe cooled: it is the differing masses of the force carriers that spoil the symmetry.

The asymmetry between electromagnetic and weak forces at low temperatures, in contrast to their symmetry at high temperatures, appears to arise for the strong nuclear force too. Careful experiments have shown that the strong force also is changing its character, becoming enfeebled relative to its character at lower energies, such as in the atomic nucleus of atoms at room temperature. The way that the fundamental quarks respond to the strong, electromagnetic and weak forces hints that the strong force is united with this "electroweak" at even higher energies. From detailed observation of the way these forces behave as the energies rise from room temperatures to those of LEP, and extrapolating onwards and upwards, theoreticians estimate that the strong force becomes part of a united trinity as energies of around 40 thousand billion billion times room temperature.

Such was the early universe: intense energy or temperature which died out, cooled, as time passed. The unification of the strong force with the electroweak at even higher energies is not yet demonstrated, as such energies are far beyond even the most optimistic design for a particle accelerator, but most theorists believe that it is likely.

It has not yet been conclusively demonstrated that these forces are all united at the extreme temperatures of the Creation, but the hints are that they were. As the universe cooled from the cauldron of the hot Big Bang, the united forces separated into the distinctive forms that science has revealed. How this happens is a key question for our time.

Supersymmetry

We have alluded to the idea that the forces of nature were originally as one but have said little about the particles on which they act. Here too there is an increasing sense of symmetry as we probe deep into the cosmic "onion," peeling away the layers that hide the true seeds of matter from our macroscopic senses. Our immediate experiences are at the outer layer of the onion, where we see matter in bulk, on a scale that is huge compared to that of individual atoms. The result is diversity. Diamonds and soot are as unlike as one could imagine and yet at the atomic layer they are both revealed to be nothing other than carbon atoms.

Within the atom there is symmetry only in the precision balancing of the electrical charges of the electrons in the outer limits of the atom and the nucleus at its heart, whereas in terms of mass they are highly asymmetric: electrons are some 2000 times lighter than the nucleus of even the lightest atom. The whole nature of that nucleus, with its massive neutrons and protons and measurable size, is quite unlike that of the electrons, which appear to be points whose size, if any, is too small for us (yet) to discern.

However, when we probe within the neutron and proton we find the kingdom of the quark. What we have called neutrons and protons are apparently nothing more than bunches of quarks gripped in tight choreography by the gluons that flit among them.

The quarks are quite similar in many respects to electrons in that they are lightweight and spin as they travel, much as electrons do. The large mass of nucleus turns out to be due to the vast energies entrapped as the fields of glue ensnare the quarks.

This symmetry is even richer. Radioactivity produces an electron accompanied by a ghostly partner known as the neutrino. The neutrino is like an electron but with no electric charge. Their response to the electroweak force shows that these two particles are in a profound way similar, symmetrical—one with electrical charge and the other with none. A similar pairing occurs with the quarks: there are two types (called "up" and "down") that help distinguish proton from neutron and that are linked by the weak force of radioactivity the same as are the electron and neutrino. So at last, at the core of the cosmic onion, we find a tantalizing glimpse of order, symmetry, sameness in the fundamental particles of matter: they are all lightweights and spin at a common rate as they travel.

There is also a deep unity among the particles that transmit the forces—the photons, gluons W and Z—in that these also spin as they travel and in a common rhythm. Theorists are currently attempting to build mathematical descriptions that contain these hints of unity; the resulting constructs are known, naturally enough, as "grand unified theories." The mathematics seems to reveal further patterns that suggest that the particles of matter, such as the quarks, electron, and neutrino, and the force carriers are themselves all related at even more extreme temperatures. Searching for this ultimate "supersymmetry," as it is known, is on the menu for experiments at CERN's new accelerator, the LHC (Large Hadron Collider, "Hadron" being a generic name for the protons and atomic nuclei which will be the beams in this new machine). The LHC, which began experiments in 2010, is the ultimate telescope capable of looking to within 10^{-12} seconds of the Creation. It is seeking glimpses of supersymmetry, to discover how it, and all symmetry, was lost as the universe cooled (of which more in the final chapter).

We have come a long way in the century since X-rays, radio-activity, and the electron were first discovered. Even within the last 50 years our perceptions of the universe have advanced profoundly. The frontier of knowledge for my parents' generation was of a highly unsymmetric universe whose seeds were the electron, proton, and neutron and a myriad of particles found in cosmic rays with little apparent rhyme or reason for them. This confusion lasted for some 20 years. The first hints of rationality came with the theoretical idea of the quarks by Gell-Mann and Zweig and was only fully clarified in the 1970s. As the cost of research rose through the 1960s and all that the particle physicists seemed to be creating was a confusing menagerie of weirdly named particles, there were some who feared that the explanations would be beyond our technological and economic means and that we would remain in darkness—that this quest for a fundamental description of our universe was doomed and would spell the ends of practical science. My generation has been privileged to see the light in that darkness, the glimpses of deep symmetry in the early universe. This is where we are today. The youngest generation will soon be seeking answers to how nature made its transition from perfection to the highly structured asymmetrical legacy that is now.

Chapter 9

Lost symmetries

In 1961, the Cuban Missile Crisis nearly led to Armageddon and the destruction of possibly the only complex system capable of knowing the "why?" of the universe. Had Leopold Lojka maintained symmetry by driving straight, or even broken it by turning left, none of this would have happened. But he turned right and with this chance asymmetry set in train a sequence of events whose consequences are still being worked out.

The Soviet leader, Nikita Kruschev, had risen to the top after surviving the tyrannies of Joseph Stalin in which millions had died. Stalin's rule went back to the Second World War: the nuclear weapons shipped across to Cuba were the result of the atomic race that had begun in that war and which had already destroyed Hiroshima and Nagasaki. Much of the technology that defines the late twentieth century has developed as a result of those events; the internet requires satellite communications which are in turn dependent on modern rockets; electronics has flowed from the development of radar; geopolitics has been inexorably determined by the devastating potential of nuclear weapons. Many of these would have eventually been developed, though the demands of war accelerated their arrival. However it is arguable that the vast technology required to make the first atomic bombs might never have happened in the absence of the Second World War.

Hitler had unleashed that war as a dire consequence of the defeat that he had experienced in the First World War in which further millions had perished. It has been estimated that during the twentieth century as many as 100 million people—that is equivalent to the total populations of Britain and France or half of the USA—died during conflicts and tyrannies that trace back to the First World War. Perhaps the world was going to fight anyway, but it is generally agreed that the singular event that set it all in train, and altered the course of history, was the assassination of the Archduke Ferdinand in Sarajevo. Which is where Leopold Lojka enters the story.

The conspirators had already tried and failed to kill the Archduke with a bomb earlier on that fateful day. One of them, a slip of a boy called Gavrilo Princip, who had been refused entry to the Serbian volunteer force, was walking down a side street returning home from the failed attempt as, unknown to him, the Archduke's open-topped car was driving along the main road a hundred metres away.

Leopold Lojka was at the wheel, driving at walking pace. It was then that he made the fatal mistake: instead of taking a left, he turned right. The Archduke, realizing the error, ordered him to stop and turn around. As Lojka wrestled with the stiff gear lever, Gavrilo Princip could not believe his good fortune—having failed earlier that day in his assassination attempt, he suddenly finds the Archduke Franz Ferdinand and Archduchess Sophie sitting in an open-top limousine, stationary in the side-street less than two metres away. Princip took out his gun, leaned on the side of the car, took aim and shot both Ferdinand and Sophie. Within a month all of the great powers were at war and the course of history changed for all time.

All of this was the result of taking a right turn instead of a left; a nugatory asymmetry can have gargantuan consequences, given a long enough timespan. With the passage of more than four billion years since the newborn earth began to cook the first complex molecules, a small trifling asymmetry at that point could have seeded the gross asymmetries of amino acids and life today.

The mythical example of lowland and highland sheep shows how animals that initially were collectively similar could evolve into distinct identifiable species; asymmetry emerging courtesy of their environment. Symmetry would imply that the sheep have legs on their left and right sides of equal length and to begin with this was indeed the case. But then evolution and survival of the fittest enters the story and over the generations causes asymmetries, favoured by the environment, to emerge. For lowlanders living on the flat, horizontal stability favours animals whose left fore or hind leg is the same length as the corresponding right one: a left–right symmetry. However, for the highlanders living on the steep mountain slopes such a symmetry would be a disadvantage: to maintain stability the sheep need to kneel on the upper side

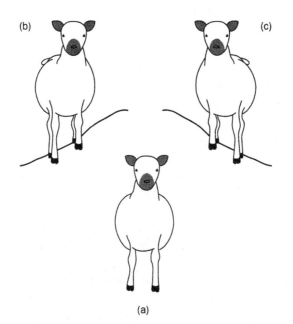

Fig. 9.1 Symmetric and asymmetric sheep: (a) shows a flatland sheep; (b) and (c) show that a mountain sheep can be a right or left leaner.

while staying upright on the lower. While this enables them to eat the grass more comfortably, it makes walking difficult.

As we saw in Chapter 3 there are always small deviations from symmetry in animals. Some sheep will have had legs on the right side slightly shorter, for others it will have been their left legs that are stunted. The former sheep will find it easier to move around the mountain clockwise, while their mirrored counterparts will move anticlockwise. In reality there are so many variables that determine a sheep's chance of survival that distortion of the legs is hardly likely to help (it is unlikely that the sheep would have left the plains and gone to the slopes in the first place). But this is a mathematician's "ideal sheep," where all the variables are neglected save one: suppose that it were the case that the legs determined survival and imagine the outcome. Whatever the mutation in the genes that causes the lopsidedness, initially there will be left leaners and right leaners. Will they be able to interbreed?

Typically left-leaning rams would mate with left-leaning ewes, and right with right. The asymmetries encoded in their respective genes will propagate and ultimately two sets of chiral sheep will emerge; mirror symmetry is preserved as there are both left and right species. However, we could imagine how this might be lost leading to only one chirality for mountain sheep.

Although the legs were, by hypothesis, the dominant genetic factor, there are other environmental asymmetries that may affect the two species. The left and the right varieties will face opposite directions on the mountain and move in contrary directions too. The two families will have different experiences, one being condemned to facing any prevailing winds or perhaps having the hot afternoon sun on its back rather than in its eyes; one or other of these may have advantages, small in detail but over long periods leading to large consequences. It is possible that a minor relative disadvantage could cause one variety to die out, leaving a complete mirror asymmetry in the highland sheep.

Analogously, it is possible that left–right asymmetry could have emerged in humans by evolutionary advantage as a small dominance of one over the other amplified with time, but this is thought to be unlikely as the chirality seems to be so general. Its origin is believed to have been universal, either in the environment as the solar system formed (for example, like the polarized light in Orion, discussed in Chapter 4) or due to intrinsic chirality in the natural forces (which we shall meet in the next chapter). My purpose here is merely to illustrate how seemingly trifling asymmetries could have immense consequences.

However, this still begs the question as to how the critical asymmetry occurred in the first place. How is it that in a seemingly symmetrical situation, asymmetry can occur and grow? This is an enigma that has intrigued philosophers for centuries. A classic example is the "paradox" of Buridan's ass.

The philosopher Buridan, in the fourteenth century, was concerned by the dilemma of free will. The standard version of the story is that he was intrigued by the perfectly symmetrical situation where an ass was standing exactly midway between two identical bunches of carrots. There is a symmetry of choice as to which of the two identical bunches it might prefer and a symmetry of information about the two in that they are equally remote from the animal; the ass is therefore unable to choose one rather than the other and so it starves. (This traditional version of the story is actually wrong! It was not an ass but a dog, and Buridan never suggests that it starves but that it chooses "at random." This actually led Buridan to develop early ideas on randomness and probability.)

In reality this does not happen. We could illustrate the enigma by drawing the ass on top of a small hill with valleys at either side, rather like the shape at the bottom of a wine bottle (the reason for this arcane choice will hopefully become clearer later!) and with the carrots placed evenly in each of the two valleys. The ass chooses one bunch of carrots or the other and breaks the symmetry arbitrarily. The influences that tip the ass to left or right may be triflingly small—a breeze turned its head perhaps—but the consequences are very different.

Suppose we sit at the left-hand side, at the bottom of the slope. If the ass comes this way, and ends up at the bottom of the hollow, we have an ass, whereas if it set off initially to the right, then we are left with a bunch of carrots. Therefore the outcome of the subtle breaking of the symmetry, the randomness of chance, depends on where you are. You end up with some experience or phenomenon that is utterly non-symmetric, even though initially the set-up was perfectly balanced.

We can suggest another "solution" to Buridan's problem by exploiting what we have learned in earlier chapters: an ass may appear superficially symmetrical but internally it has mirror asymmetries in its molecules, its brain, and a consequent preference for right or left. Leopold Lojka happened to choose a right turn; he could as easily have chosen to go left, but the consequences were singular. The ass will choose one or the other, it does not matter which: either left or right, the breaking of symmetry marks the difference between starvation and staying alive.

Fig. 9.2 The dilemma of Buridan's ass.

Another example of symmetry that "must" be broken if any progress is to happen occurs often in restaurants or at banquets where several are seated at the same table. A banquet can be a high-risk occasion if the guests are seated at long, straight tables, as for the duration of the meal your opportunities for conversation will be dominated by the individuals to either side of you—if they are bores, your evening could be ruined. It can be particularly galling if you are at a table containing an odd number of places as pairwise conversations leave at least one person in isolation. Expert diners choose to avoid the ends of such tables, failing which they ensure that they speak to their neighbour immediately, thereby defining the wave of pairs that spreads along the table. The odd one out may be somewhere near the middle, in which case there are two (or more) domains of conversing diners. (Phenomena that we shall meet later, and which arise in the structures of magnets, or perhaps even the large-scale structure of the universe, have their analogues in more homely examples such as these.) Far better are those occasions where the diners are seated at circular tables enabling a variety of conversations to take place, even involving the entire table. This situation is also guaranteed to break the ice at the start of the meal as all members are forced to interact as follows.

Visualize the scene. You are one of maybe eight people all sitting around the circular table. Let's suppose that the occasion is very important so that the tables have been laid with special care. Viewed from directly above the scene will consist of 8 heads at 8 places, laid out in perfect symmetry with knives, forks, and place-mats in front of each. Your immediate neighbours are equidistant from you, as are all of the juxtaposed places around the circle. Midway between each place-mat is a table napkin; 8 in all. The waiters deliver the soup. Everyone is very polite and waits until all are served. Now comes the moment of decision as you have to take a napkin to protect your expensive dress or dinner jacket: is your personal napkin the one to your right or to your left?

Fig. 9.3 The dilemma of the napkins: is the napkin on the right or the left the one to choose?

Elsewhere on the table someone randomly chooses their napkin, say the one on their left. Immediately this defines which napkin belongs to you and which to your partner. Once someone makes the choice, that is it: the symmetry is broken and everyone is forced to go the same way. What was originally a symmetric situation of plates and napkins to either side of you has become asymmetric as all reach to their left (or to their right).

This example also illustrates the "domain" idea that we met earlier. A problem will arise if two of the diners who are not adjacent to one another should choose differently, one picking the napkin to their right while the other selects the left. The guests immediately adjacent to each of them will follow their example and pick to the right or left accordingly. The left and right "waves" of napkin pickers will spread around the table until the two waves meet. On one side of the table a guest will have no napkin at all, while on the other side there will be a spare one. Two "domains" of diners ready to eat, clothed with napkins, will result in what is known as a "defect" at the interface between them.

A simple demonstration of this effect in a real physical situation consists of a set of small compass needles, each of which is suspended at its midpoint and arranged in a matrix. First you set them whirling around, which you can do by holding a larger and more powerful magnet at one side and then rapidly moving it from one side of the matrix to the other. All of the compass needles will start rotating several times each second, pointing at every direction around the circle as they go. In this agitated state their orientations are spread uniformly, symmetrically, around the circle. Gradually they slow down. One or other of them slows sooner and stops at random, its magnetic influence pulling its neighbours into line; elsewhere another needle will have slowed and ensnared those in its vicinity. Like the diners at the table, this set of neighbours may have chosen the same orientation as those on the other side of the matrix, or different. In the latter case, domains of aligned compass needles will arise. There is a symmetry, a pattern, among the needles but it now consists of structures rather than the uniform circular symmetry of the start.

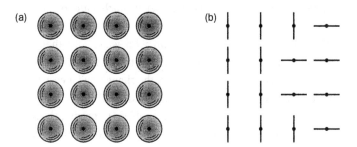

Fig. 9.4 The matrix of compass needles: in (a) they are rotating fast and all directions are used by the magnets; in (b) they have slowed to rest and their mutual magnetism grips them into domains.

This example involves a change in symmetry as the energy falls: when rotating, the magnets have energy in their motion—kinetic energy—which is lost as they come to rest (it has been dissipated as heat and in the magnetic fields that invisibly link the magnets). So the former, high-energy state is uniformly oriented, whereas the final low-energy state has structure. We shall meet further examples of this later on. One of the questions that we need to answer is why it is that some situations stay symmetric whereas others can change.

Spontaneous symmetry breaking

Stars are spherically symmetric because the force of gravity that forms them cares only about the radial distance between two mutually attracting masses and not for their relative orientation. Gravity is the cause of the star and gravity is spherically symmetric; the star that results is, in consequence, spherically symmetric. Pierre Curie, husband of Marie Curie, realized the importance of symmetry over a hundred years ago and put the above into a formal principle: if some phenomenon gives rise to some particular effect, then the symmetries of the phenomenon will appear in the effects that it generates.

Curie's principle is true but it is temptingly easy to go one step further and get into trouble. To illustrate the problem, consider a simple case of perfect circular symmetry which suddenly appears to "go wrong."

Fill a circular saucer to the brim with water and then drop something into it. There will be a splash, the liquid will be spilled, and where it ends up will depend on the shape of whatever fell in, where it landed in the saucer, the direction that it was moving as it hit, and so on. It would be a hard problem to predict the shape of the splash in general but in one case it should be simple. The special case is one of perfect symmetry.

Suppose that the falling object is perfectly circular, that it hits the surface of the liquid precisely at the centre while falling vertically with no motion horizontally. Take a video of the action and freeze the frame at the instant that the object touches the surface and begins to displace the liquid. We have here a situation with perfect symmetry: in each and every horizontal direction the system looks the same, the distance from the edge of the object to the edge of the saucer is the same, and the liquid is being displaced the same at each and every point around the circular edge of the invader. The liquid is in the first stages of making a splash: the cause of the splash has circular symmetry and so the effect, the splash, should also manifest circular symmetry according to "common sense" or to Curie's principle.

If you were to view the video one frame at a time you would indeed see that a circular splash arcs upwards from the impact site initially. Just as you are about to congratulate yourself on having used symmetry to predict successfully the outcome, you suddenly notice in the frozen images of the video that although the base of the splash is a perfect circle, the upper edge of the rising liquid has broken into several individual spikes which in turn have thrown off minute droplets, evenly spaced around the circle such that the whole looks like a circular coronet adorned with individual gems. As these fall back to ground, landing on the dry paper that you thoughtfully surrounded the experiment with, you will see that the splash wets the paper at discrete spots where the individual droplets land while remaining dry at the intervening gaps.

We have started off with perfect circular symmetry in the liquid, the bowl, and the falling object, but the effect, the splash, is not circularly symmetric. It manifests symmetry, to be sure, in that the drops are likely to be evenly spaced all around the circle but there is no circular symmetry overall. For example, imagine that there are 12 drops which fall around the circle like the 12 segments of a clock face. If you choose the directions 5, 10, 15, and 20 minutes past the hour it is wet; but at 2, 7, 12, and 17 minutes past the hour the hands would be pointing at dry spots. So instead of circularly symmetric dampness we have a different symmetry consisting of discrete dry and wet regions interspersed.

Something appears to be wrong with Curie's principle; somehow the symmetry has changed from one form to another. In modern scientific jargon one would say that the symmetry has become "spontaneously broken" or alternatively that the original symmetry has become "hidden."

The phenomenon of hidden symmetry is so common that it is perhaps surprising that only in recent years has it taken centre stage in some areas of science. When its effects are first encountered, as in the case of the water drop, it appears almost magical. How and why does the symmetry change? When is Curie's principle true and when do we have to be careful?

Some common examples of changed symmetry

You have loaded up your shopping trolley in the supermarket, or piled your suitcases on to a luggage cart at the airport, and then you try to push the thing forwards. It seems to take on a mind of its own as the wheels flick from one direction to another but never, or at best rarely, in the forwards direction that you are trying to go. You are experiencing spontaneous breaking of symmetry.

First, what is the symmetry? Suppose that the trolley is perfectly engineered so that its left- and right-hand side, when viewed from your position at the rear of the trolley, are mirror images of one another. Place your left hand on the pushbar at some distance to the left of the mirror axis and your right hand the same to the right of it. Line the wheels up so that they are pointing precisely forwards. What we have is a mirror symmetry from front to back along the middle of the trolley.

Now start to push it evenly with left and right hands. The symmetry of the situation would suggest that the trolley will move forwards along the line of the mirror. Briefly it does but then the front wheels start to flick wildly from side to side until both are pointing one way or the other but not straight ahead. The trolley twists to left or right but not where you want to go. The collection of objects—namely you, the trolley, and the floor—may all be perfectly symmetric but the outcome is not. We say that the "system" (the objects) is symmetric but the "solution" (the outcome) is not.

A symmetric system will have symmetric solutions only *if these are the most stable possible*. However, if they are unstable, the system will do something else. Pushing the trolley forwards on those little wheels is an unstable situation. It does not take much force to deflect a wheel from its centre line, especially on the smooth floors in supermarkets. Were you to be pushing it on gravel the force required to move the trolley forwards, or to deflect the wheels sideways, would be much larger; moving your belongings would be hard work but at least the wheels would stay straight.

A more banal example, though one where it is easier to see where the "magic" happens, is if you suspend a snooker cue from a beam, such that the cue's tip hangs centrally above a static roulette wheel. Having done so, cut the suspension and the cue will fall by chance onto red or blue; when suspended, all directions appeared to be equally symmetrical, any direction of fall as likely as another. This is the principle behind the more sophisticated roulette wheel. In roulette you feel that if you knew the speed of the ball, the wheel, and solved Newton's equations of motion, then you could beat the casino. (Some students tried this in Las Vegas and told the story in *"The Newtonian Casino."*) The case of the snooker cue is more subtle: cutting the string already breaks the symmetry—which direction were you standing when you cut? Ultimately, no snooker cue is perfectly circular in cross-section; at the molecular level there are dips and bumps that break the symmetry.

Similar to the snooker cue example, we could imagine in Fig. 9.2 that Buridan's ass was replaced by a ball sitting precisely on top of the symmetrical hill. The spot where the ball rests is "flat," if only for the microscopic distance where the ball and ground are in contact. As such the ball is stable and could remain there forever. However, the slightest disturbance will displace the ball fractionally from this perfect spot. Its moment of glory on its pedestal is gone and from that instant it will be literally downhill all the way until the ball comes to rest in the hollow. As for the snooker cue, so for the ball: small imperfections on the surface will break the ideal symmetry and prevent it staying atop the hill in a position of "metastability." The hill top is not the most stable position; the hollows, however, are.

Another example of a small asymmetry having large effects, that you can try for yourself, involves an empty soft drinks can. Such cans are flimsy and can easily be crushed in your hand. However, it is possible for the empty can to support the weight of a human (so long as you stand carefully balanced with your centre of gravity over the middle of the can's top). The reason is that symmetry all around the circular top spreads your weight evenly and no one spot chooses to give way more than any other. But the smallest imperfection around that circle will cause a catastrophic crushing of the can (and this is why the experiment is hazardous for a human and safer with an inanimate heavy weight instead). Having balanced the weight atop the empty can, a small pressure on the side wall of the can will cause it suddenly to collapse: the weight crashes to the ground. This is an example of falling from a state of high potential energy (the weight supported on the top of the can) to lower energy (the weight on the floor). This is a metaphor for systems seeking states of lower energy and the symmetry changing as a result. It also illustrates that when even a triflingly small asymmetry enters the scene, change can happen with startling suddenness.

In the example of the can we introduced an asymmetry (the imperfection in the side wall of the can) in order to initiate the collapse. For the snooker cue and the shopping trolley we tried to be extremely careful so as not to spoil the symmetry, but even so the symmetry disappeared, as in the case of the water splash earlier in this chapter. The reality is that systems are never perfectly symmetrical. They may appear so at a gross scale but under a microscope imperfections will appear. The machining of the wheels in mass production factories will introduce small flaws; wheels will have been bent slightly as they hit other trolleys or the walls and

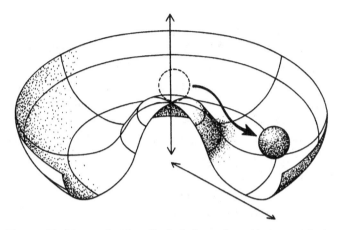

Fig. 9.5 The Mexican hat. Initially the ball is at the stable point at the top of the hill: it has high potential energy. The potential energy is lowest in the rim at the bottom. The ball falls into the rim at random, breaking the circular symmetry.

doors of the store; your goods are not evenly distributed; and so forth. Suppose that you are careful to ensure that none of this happens; imagine setting up an ideal environment and even have a machine push the trolley so as to avoid your innate tendency to push one hand more powerfully than with the other—what then? Even in this carefully prepared symmetrical experiment, spontaneous symmetry breaking will occur as subtle imperfections arise at the molecular level. The wheels are built from molecules of plastics or other preservatives; these involve complex chains with their own asymmetries.

Well, perhaps we could overcome that, at least in principle. Suppose that we prepare a special wheel surface from some ideal spherical molecules which will guide us across the ultimate smooth horizontal floor, or, in the example of the ball in Fig. 9.5, engineer the perfect sphere, placed with perfect precision on top of the hill: what can go wrong with this? The answer is that the warmth of the room will defeat us. The thermal agitation of the molecules as they vibrate and rotate, bouncing one from another while still glued as a whole, will introduce randomness and instability. A small random disturbance, metaphorically turning one molecule to the left rather than the right, will grow until the system finds a stable configuration.

Perhaps we could imagine being at a temperature of absolute zero where all molecules have come to a halt; could we then beat asymmetry? Even here nature conspires to defeat us. There are subtle quantum effects that prevent you positioning anything with perfect precision while at the same time keeping it perfectly still. Conversely, if an atom is brought to rest, its precise location is inherently unknown. I cannot explain why this is—it is in the fabric of space and time of the material universe. The consequence is that an atom at rest cannot also be at the precise spot atop the hill and so will begin to roll down the slope at random. The bottom line, both metaphorically and literally, is that the metastability atop the hill will be lost and all things will fall into the circular trough.

Freezing: from symmetry to structure

So much for shopping trolleys, snooker cues, balls on hills, and drinks cans; what do these have to do with the real world? The first example that I will look at is the dramatic change that occurs when the temperature falls below freezing point. The beautiful, highly structured form of the snow flake somehow emerges like a butterfly from the bland uniformity of water and vapour. Current thought is that the structure in the universe today emerged from an original symmetry when the universe "froze."

We started our quest in Chapter 2 by noting that a sphere is the natural shape when the force attracting the constituents cares only about their distance apart but not the direction, and we cited gravity as an example. A critical part of this is the fact that gravity attracts everything to everything else; there is no repulsive force at work—this is an essential reason for the spherical symmetry that results. By contrast, when both attraction and repulsion are at work, asymmetry tends to result; in such cases instabilities can arise and the overall symmetry change. For the ball atop the hill in Fig. 9.5, gravity of the earth attracts it down while the solidity of the hill prevents it making a direct drop.

The competition between attraction and repulsion at the molecular and atomic level can also cause asymmetry. Here it is the electromagnetic forces, with attractions among atoms at moderate distances but repulsions as the electrons of neighbours get too close to one another, that is responsible. The fact that symmetries can change when repulsive effects are present is the cause of many well-known phenomena and I will illustrate these familiar cases before entering the atomic world.

An example on the motorway is where many cars are all travelling at the same speed in the same direction. If there are few of them they maintain their distance and there is symmetry along the road: a car passes every ten seconds say. Now imagine that it is rush hour and that more cars have entered the motorway. Cars are now passing every five seconds, then every four. At this point the symmetry changes and instead of a smooth flow of traffic there will be bunches as cars slow and gradually work their way to the front of the queue and speed off again, only to meet another blockage further along the road: the symmetry has changed. Instead of regular, temporal symmetry (one car every 5 seconds), we have a clump every mile separated by thin regions in between. There is still a symmetry viewed from the traffic helicopter as it travels along the route but it is different from the original. What is happening?

There is a "repulsive" effect that disturbs the flow and unbalances the system; it is the amount of time that it takes drivers to react. If you get too near to the car in front you brake in order to "repel" yourself from it. The car behind brakes more sharply and the whole queue rapidly comes to a halt. The cars at the front can accelerate off into the open road, soon to be followed by those behind them; meanwhile cars further back are slowing down. This change of configuration, from smooth flow to clumps, all began because one motorist moved marginally nearer to the car in front than reactions would allow.

There is another way of preserving the ideal flow and that is if all the cars were to slow uniformly to a lower speed. In this solution they can travel nearer to one another while their drivers are still able to react and thereby remove the "repulsive" effect of the drivers' reaction times. The net result, paradoxically, is that more cars can pass along the road per minute while travelling slowly than in the former, high-speed case. This mathematical truth was realized only recently in Britain with the introduction of speed restrictions on London's orbital motorway, the M25, when traffic density becomes too high. The alternative solution in some parts of the USA has been to restrict entry to the freeway in order to help maintain the high-speed symmetric ideal for those that are already there.

Another example will be familiar to skiers. After a fall of snow everyone rushes out to the slopes. The snow is smooth, the piste is symmetric but for the direction of its slope. This trivial breaking of symmetry defines which way the skiers go—downwards—but if the slope is uniform there is nothing that would appear to distinguish one part of it from another. The pristine slope is symmetric with a uniformly smooth surface along its length. However, if the slope is steep enough, within a few hours of intensive skiing this symmetry will have disappeared as hillocks, known as moguls, form; if there are no more falls of snow, within a day or two these moguls will have become so large as to be challenges for advanced skiers to enjoy.

The origin once more is instability; the "repulsive force" is the fear of the skiers who are unwilling to ski ever faster under the pull of gravity and so brake by swerving from side to side on the edges of their skis. Small ripples develop which cause later skiers to make their turns at the same place, eventually turning the ripples into towering moguls. By this stage even the best skiers are forced to slip and slide down the faces of the moguls, shifting the snow from the edge to the base. The effect is that the individual moguls gradually move up the hillside becoming ever more pronounced as their fronts are sharpened and their peaks are replaced by snow chipped from the next mogul up the mountain. This too is analogous to the traffic jams on the motorway, where you enter at the rear and work your way to the front as part of an overall symmetric structure that is on the move.

This example also shows another feature of asymmetry: the notion of a critical value in some variable quantity that determines whether symmetry survives or when the asymmetry sets in. On the gentle slopes—the "green pistes"—the smooth symmetry of the snow is preserved and moguls rarely if ever occur; the reason is that beginners are skiing too slowly to leave a lasting impression, while advanced skiers speed straight through without exceeding the "fear factor." However, as the slope steepens there comes a point where fear takes over, the skiers start to swerve and the moguls begin to form. It is the steepness of the slope, the increasing competition between the downward pull of gravity and the frictional drag of the snow, that determine when the moguls and the asymmetry turn on. The critical value is somewhere in the red to black piste grading; below this (blue and green pistes) sharp moguls tend not to form, whereas above it (the steeper parts of red and commonly black pistes) the asymmetry of moguls takes over. A more common example of such a variable is temperature, and the critical values are where the system changes from one "phase" to another, as in condensation or freezing.

Having persevered with all of this preamble, let's now take the plunge, or at least dip our toes in the water.

The frozen universe

When gases or liquids are cooled we are all familiar with them becoming liquid or solid respectively. These are examples of what is known as a change of phase, which arises when the component parts of the system remain the same but are redistributed in new characteristic ways.

The hotter a substance is, the faster its molecules move and the more disordered their movements are. In a gas, the molecules are far apart and move essentially independently of one another except when they make chance collisions; the gas occupies all three dimensions of its container and no one direction is favoured over another (I am imagining that we are in a space station away from the gravitational influence of earth) and spherical symmetry is the rule. You could in principle swap two molecules and the gas would remain the same.

Now cool the gas. Temperature is a manifestation of agitation of the individual atoms and molecular constituents. As the temperature falls so the molecules slow their frantic dance, and their mutual electromagnetic attractions pull them to one another, as was the case of the whirling compass needles in Fig 9.4. As the molecules move ever closer to one another, their original spheres of influence begin to shrink. Having started with spherical symmetry and with nothing to single out one direction over another, it would be natural to expect that the spherical symmetry will survive as they cool. However, this is not the case; something akin to the shopping trolley instability begins to take over.

Imagine for a moment two atoms getting very close and trying to get on top of one another. This cannot happen. Why it cannot is due to a profound property of quantum mechanics but that they cannot is apparent from the fact that you are sitting comfortably, I hope, and not sinking through the seat or the floor below. Make your hand into a fist and rap it on the table. It did not pass through, yet the atoms in the table and in your hand are mostly empty space. Electrons in the outer reaches of the atoms in the table and in your hand are trying to occupy the same space but mutually repel each other—solidity results.

A similar effect occurs when the gas cools to liquid, or further to solid: molecules that were originally almost free agents are increasingly meeting and influencing one another, their constituents trying to occupy the same piece of space as the whole cools and contracts. The ideal situation where we started with a sphere of atoms in a gas that try to condense to a sphere of liquid or solid with all atoms ultimately at a point, does not happen. Their mutual repulsion as they encroach creates an instability in the system. The smallest random disturbance, where two atoms meet slightly off centre, will have escalating effects and the system will find a more stable configuration whose overall symmetry may differ from that of a sphere.

Such is the case for water at room temperature as it cools to and below 0°C. If now we focus on the water molecules we can begin to understand the nature of the phase change. Above 0°C, the water molecules are colliding so often that no structure can survive; in effect the motions are random, filling all three dimensions symmetrically. Suppose now we put a small piece of ice into the liquid. The fast-moving molecules in the warm liquid hit the surface of the ice, the violence of the collisions disrupt the locked-in molecules of the ice: the ice "melts." However, if the temperature of the water had been much cooler, the violent motions of its molecules would have been less and their impact on the ice correspondingly reduced. If the temperature of the water had itself been 0°C, its molecules would no longer have had enough energy to disrupt the molecules in the ice and the ice would remain frozen; and if no energy was supplied to the water such that its temperature remained at 0°C, the entire contents would freeze.

Left to their own devices, physical systems will eventually fall into the state with the lowest energy. At 0°C, when water freezes, it gives up energy as heat even though the temperature stays at freezing point; conversely, when ice melts it absorbs energy. Ice at 0°C contains less energy than liquid water at that temperature; water in its lowest energy state at 0°C is solid ice rather than liquid.

In the liquid phase, the molecules of water are moving at random; in ice they merely oscillate about stationary positions. Water molecules associate strongly to one another. In ice crystals this association is a highly ordered but loose hexagonal structure, recognizable by the beauty of snowflakes. When ice melts this orderly arrangement breaks down and the molecules pack more closely together. This makes cold liquid water denser than ice, which is why the deceptively solid surface of frozen ponds is merely ice that is floating on top of liquid water and such a danger to the unwary.

It is the shapes and electronic linkages between the hydrogen and oxygen atoms in the frozen water molecules that determine the locked-in pattern that we see in ice crystals. A picture of a snowflake shows this beautiful six-fold symmetry. The orientation of an individual flake could be in any direction, like the drops in the water splash that we met earlier. So the snowflake represents a continuous breaking of symmetry; the initial point could have been anywhere around the circle but once the freezing starts at one spot, then, like the napkins at the circular table, all the positions of the other members of the sextet are determined.

The snowflake is one visual example of the breaking of a continuous symmetry as water moves from the liquid to the solid phase. There can be changes of phase, and symmetry, within the solid state alone, as in the case of magnetism. The directions of north and south poles of a magnet are determined by the rotary motions of individual atomic constituents all dancing as a team.

Fig. 9.6 The six-fold symmetry of a snowflake.

If the dance troop are rotating clockwise the north pole is in a certain direction, whereas if they reverse and go through their routine anticlockwise, it will be the south pole instead of the north that points the way. When a magnet is heated, the dancing constituents become like whirling dervishes and their motions are so randomized that all cancel out; all memory of north and south is lost. When it is cooled again one finds that a group of dancers in one region have taken their partners and are all dancing clockwise, while on another part of the floor, initially unaware of their far-flung friends, the dance partners have by chance started gyrating in the opposite direction. There are regions where north poles are all aligned; there are other regions where it is the south poles that are oriented like that. These separate regions of magnetism are called domains: it is the result of order being imposed on unruly components who take their cue from their nearest neighbours.

These domains in the structure of magnets are physical examples of the human interaction at the banquet table or, more directly, of the matrix of compass needles in Fig 9.4. It is possible that some of the asymmetries that are familiar to us in the world at large, such as the characteristic spiralling of DNA among others, could be of the domain variety in the sense that chance caused left-handed preference in one region, right-handed in another, and over the ages natural selection chose a winner.

Lost symmetry (often called "hidden symmetry" or "spontaneously broke symmetry") is being increasingly recognized as pervading many physical phenomena. Following seminal ideas by the Japanese-American physicist Yoichiro Nambu around 1960, theorists suspect that it is responsible for the emergence of a structured, cold universe. As stated earlier, we believe that the original, hot Big Bang was initially symmetric; this may be a comment on our limited imagination or it may be the reality. In any event, that is our best starting point given present knowledge. Initially the universe was hotter than any star is now and immeasurably hot compared to the ambient temperature of -270°C of deep space "today." As the universe cooled from the cauldron of the hot Big

Bang, various rubicons had to be crossed as certain critical temperatures were reached. Successive moments arrived when nature underwent changes of the phase and had to decide which way to break the symmetry, as at the temperature where electromagnetic and weak forces separated from the hot unified force field.

Exactly how this occurred is still being argued about but there is general agreement that it is connected with the transition from hot to cold. It is becoming clear that not only water and hot metal change their microscopic symmetries as they cool, changing phase to snowflake or magnetism such that the original symmetry becomes hidden, but that the forces of nature and the pattern of elementary particles, which form the fabric of the material universe, also have done so. Change of phase and change in symmetry go together, so we suspect that the universe must have undergone a phase change.

In Chapter 13 we will encounter the bizarre concept of the "Higgs field" that is believed to pervade all of space, having condensed as the universe cooled below some $10^{17\circ}$ about 10^{-14} seconds after its birth. Theorists suspect that the original Creation was indeed perfect and that the true symmetries of nature have been hidden by that momentous phase transition, since when the Higgs field has pervaded the universe. This could turn out to have been the most profound and far-reaching of all phase transitions—the act that hid the original symmetry of the Creation from view. One of the questions that is interesting scientists is whether the existence of life also is due to this phenomenon.

Chapter 10

Nature's sleight of hand

Had it been possible to have placed bets at the dawn of time, what odds would have been offered against the emergence of life? Fifteen thousand million years later, we are here—complex systems of molecules that are self-aware.

In seeking the origins of the asymmetries that appear to be essential for existence and reproduction, there are three distinct phases into which we can divide the history of time. The first of these, the "cosmic era," lasted for less than a microsecond following the Creation. It was in this brief moment that the elemental particles, the electrons, and quarks, emerged victorious and the forces that would fuse them into more complex structures began their work. It took 100,000 years of cooling for the quarks to condense into the protons and neutrons of atomic nuclei and then to combine with electrons to make the first atomic chemicals and compounds. Once these were in place the second stage, the "chemical era," began. With simple atomic elements as the ingredients, the molecules of future life began to be "cooked."

Several thousand million years were to elapse before the sun and earth were formed, so it is possible that the molecular seeds of life could have been made long before the planets existed. One theory is that comets and other cosmic debris, smashing into the atmosphere, brought with them the primitive organic molecules that are our ancestors. Alternatively, they may have been fused only in the newborn earth.

Once the earth was formed, complete with these organic templates, the third phase, the "biological era" in which we now find ourselves, began. For some four billion years, nucleic acid polymers have replicated and proteins been synthesized such that life as we recognize it has evolved. The shapes of these complex molecules that entwine to make the amino acids, the templates for DNA, and life, naturally connect like the clockwise or anticlockwise twists of a spiral staircase, as we saw in Chapter 4. Reproduction is more efficient when there is a chiral asymmetry, with one direction of the twist dominating. Once living processes had selected handedness, protein synthesis and the chiral selectivity of enzymes could have ensured that such handedness was passed on through the generations.

As we discussed earlier, the resulting asymmetries in the molecules of life could have been generated entirely during the biological era. However, the discovery of left-handed "standard" amino acids on the Murchison meteorite (see page 76) suggests that the chirality of these molecules was established before life on earth had started and was in fact already present in the chemical era. Did it arise during the chemical era, and what can we say about potential culprits for having initiated this asymmetry? Or was it instead imprinted into the fabric of matter as it emerged from the Creation in the cosmic era? These are the questions that this chapter will address.

On the trail of cosmic asymmetry and the origins of life

We are made of stardust. Some five billion years ago, carbon, nitrogen, oxygen, and other elements, "cooked" in stars, ejected by exploding supernovae, were entrapped in the gravity of the newborn earth, making the primeval soup from which life would eventually emerge. The atmosphere then was hot and sticky, supersaturated with molten mess much of which eventually cooled into the rock beneath our feet. In such a "soup" there are complex turbulences but over time the denser substances sink to the bottom while the less dense float upwards. So gravitation breaks the symmetry, differentiating the ideal symmetrical uniform mix into discrete components. The question is whether such an asymmetry could disturb the cookery in the prebiotic soup and lead to a chirality in the molecules.

A current is basically anything that flows. As currents of water flow from high ground to low in the earth's gravitational field, so electric current flows from regions of high electrical potential to those that are lower. When different substances are brought into contact, electric current may begin to flow, as in a battery.

In the primeval atmosphere, where assorted mixtures of chemicals were sinking and floating in and out of contact, electric currents would have been common. It is in the flow of these currents that asymmetry sneaks in. The positive and negative electric charges in atoms are balanced in magnitudes but asymmetric in where they are and how they behave. The positive charges reside in the bulky, static, central nucleus, whereas their negative counterparts are the flighty, peripheral, lightweight electrons. It is these negative charges that flow and form the electric current; the positive charges stay put.

Currents are in motion inside the earth also. As our planet spins on its axis once every 24 hours, its surface rushing eastwards towards sunrise, the ensuing rotary motions of electric charges within its molten core create a magnetic field. The earth becomes a natural magnet, a property that has been used for navigation ever since the invention of the magnetic compass.

The magnetic field of the earth attracted and repelled the wild electric currents in the prebiotic soup. Their disturbed motions in turn created further magnetic fields that would affect other nearby charges. These magnetic forces helped to weld molecules such as the amino acids that, over the aeons, built the more complex systems of life.

Even the biggest computers cannot detail the whole sequence of interactions that led to the formation of complex molecules, but there is one thing that we can be sure of: if we were to view the sequence in a mirror it would show a different set of events than happens on earth. These magnetic effects can give a distinct handedness or chirality that ultimately is traced back to the "accident" that electrons, negatively charged, are light while the positive protons are bulky. Had it been positive charge rather than negative that had set up the original electric currents in the primeval soup, or had the earth rotated in the opposite direction, the resulting motions would have been the mirror image of what actually happened. So although the basic electric and magnetic forces are mirror symmetric, the end products can be mirror asymmetric due to the naturally occurring asymmetries—the matter—antimatter asymmetry in the universe (in the sense that the lightweight electrons are negatively charged); gravity differentiating otherwise uniform mixes of atoms; and the rotating earth creating a huge magnet.

Although this may have led to chiral asymmetries in molecules, that have been amplified and perpetuated by reproduction and natural selection, this seems unlikely to be the source of life's asymmetry if the evidence of the Murchison meteorite (see page 76) is a true guide. The received opinion is that some, perhaps most, of the organic molecules that are life's templates originated in extraterrestrial environments and fell to the earth during the period, some 4 to 4.5 billion years ago, when the young planet was being intensely bombarded by carbonaceous asteroids and comets. It has been estimated that some 1,000 tonnes (a million kilograms) impacted the earth every year for a period of 100 million years, and could be a substantial proportion of the total terrestrial biomass today. Furthermore, it seems likely that amino acids could withstand the impact, remaining stable and intact. If the molecules of this material were already mirror asymmetric and maintained their chirality during impact, then at this juncture life could have started and the chirality been perpetuated.

This changes the question of "how did the asymmetries of life begin?" to "what is the extraterrestrial environment created by the molecular asymmetry?" A possible culprit emerged recently (see page 77) when astronomers discovered that in the Orion nebula, where new stars and also organic molecules are forming, there are large amounts of electromagnetic radiation that is polarized anticlockwise. There is enough energy in this light to break up molecules, and the fact that the light is polarized will affect the left- and right-handed versions of the molecules differently. Thus the polarized light can destroy one of the chiral forms while leaving the other intact, with the result that molecules in the Orion nebula are likely to be, or become, mirror symmetric. If what is happening in Orion "now" is an example of what took place as our solar system was forming five billion years ago, it could mean that our DNA today is a fossil remnant of an intense polarized light that shone long ago.

All that needs to be explained in this scenario is how the light came to be polarized. A probable answer lies in the examples of lost symmetry in Chapter 9. As the stars form, the individual pieces of hydrogen fall on top of one another under their mutual gravity and maintain the spherical symmetry. However, small instabilities occur as billions of separate stars are born and begin to orbit around one another. (We see this in the rotation of galaxies or even of the planets around our sun.) Magnetic fields, caused by the swirling and rotating electrical charges in the plasma of the stars and interstellar space, disturb the rippling electromagnetic fields that form the lightwaves. The light becomes polarized. Nature does not care whether this polarization is clockwise or anticlockwise but once started, the asymmetry propagates, as in the case of the napkin selections at the banquet (see page 166). In that example the choice could be different from one region to the other; such an effect can happen in the universe at large. In some regions of space (and here "region" can easily cover distances far larger than an individual solar system) one or other polarization can dominate and the preferential destruction of one handedness of organic molecules begin. If this is the cause of the chirality of our life forms, then all amino acids in our solar system, on Mars, asteroids, and comets, should have the same handedness. By contrast, if some day we find such molecules in the remote regions of interstellar space, their handedness could be the same or differ from ours.

So much for the biotic and chemical eras as candidates for the source of life's asymmetry; could the origin have been imprinted even earlier, in the cosmic era? In 1956 scientists discovered that nature can distinguish absolutely between left and right in certain atomic processes. In effect, there are processes that can occur in the mirror that have no place in the real world. Recently some scientists have been asking whether this intrinsic chirality in the laws of nature could have led to the mirror asymmetries of life.

The left hand of creation

When Henri Becquerel discovered radioactivity, he had not only found the means to expose the deep structure within atoms but unknowingly stumbled on the process that reveals the sinister nature of creation. Sixty years later, with the more sophisticated technology available, physicists discovered that Becquerel's radioactivity distinguishes absolutely between the real world and that behind a mirror. It was as if you raised your right hand and your image in the mirror raised the wrong one.

This is exceedingly profound. To understand why, and to appreciate the consequences, let's return to the looking-glass world that we met in Chapter 3. Recall first how we exposed the illusion of mirror symmetry by imagining ourselves at a fancy dress party where the guests were dressed as harlequins. The conditions for admission were that one's costume was red on the right and lilac on the left. In the mirror one sees someone who does not exist at the party—a harlequin with red on the left and lilac on the right. If in reality all of us were harlequins, it would be possible to tell whether a photograph pictured the real world or its mirror image. The question at issue is this: is there anything in nature that is like the harlequins and that can distinguish reality from its mirror image?

Given the number of examples of left–right asymmetry that we met in the earlier chapters of the book, you might think it would be easy to tell whether you were looking at a real event or its mirror image. In practice it is not so simple.

Imagine a room, one of whose walls is a mirror stretching from floor to ceiling and from one end of the room to the other. The impression will be that the room is twice as large and that whatever you do in its "real half," your enantiomorph will faithfully execute in the mirror half. Can the mirror image do anything that cannot occur in the real room?

Such rooms do exist, most commonly used by ballet dancers who perfect their art by observing their own performance. The effect is as if there is another room beyond the real one in which dancers are faithfully performing the mirror image routines: as the real dancers pirouette clockwise in the dance hall, the mirror room beyond shows a similar group twirling anticlockwise. Clockwise or anticlockwise the dervishes can whirl and there is no way to distinguish the reality from the mirror image in a photo taken of such a dance studio.

However, there are ways that one might guess. Were the room to be used for a written examination rather than a dance, then the dominance of right-handers over left in one image would distinguish the reality from the mirror. But could you be sure? You might for example have been deceived by watching a group that had been hired from a theatrical agency specializing in left-handers. More realistically, were you to see only one person then it would not be possible to tell whether you were viewing reality or its mirror image for, as we have seen in Chapter 3, one in ten people is left-handed.

Were we to look beneath the surface then we could make a better guess as our internal organs are asymmetric left and right, but even here there are occasional individuals that have emerged mirror-reflected. However, if we were able to examine the spiralling of their amino acids, then we would have a more certain guide as to whether we were seeing the real thing or its mirror image.

The DNA of all humans coils the same way and as such enables an absolute distinction between reality and the mirror reflection.

While amino acids in living things coil one way, there is nothing so far as we know that forbids the existence of their mirror images. Looking at the picture of some amino acid in isolation, could we tell if it was from human tissue or a mirror image of the enantiomorph created in a laboratory? Presumably not.

Until 1956 it had been generally accepted that the deep underlying laws are unable to distinguish between real events and those behind a mirror. Mirror symmetry had been thought to be a general truth, a perfection in the laws of physics, such that whatever happened in a right-hand sense could equally well occur left-handed, in which case the world behind the mirror would be as real as that in which we find ourselves. This all changed with the discovery that deep within the atomic nucleus the force that causes radioactivity exhibits a chirality in that the mirror image of Becquerel's process shows a sequence of events that do not occur in the real world.

The radioactivity that Becquerel had discovered involved the transmutation of the elements. Inside the atomic nucleus a force was slowly destroying neutrons or protons, converting one into the other. This force is so feeble in normal conditions that it is known as the "weak force." We can see it at work every day as the sun shines (page 142). The warmth that we, our ancestors, dinosaurs, sea creatures, and the first amoebae all basked in comes courtesy of this weak transmutation. The sun has now used up half of its original protons; it is halfway through its life. This feeble insidious force is where nature has hidden its sinister secret; the sun in the mirror is not the same as that we see through the window.

The historic experiment that first uncovered mirror asymmetry did not use the sun but instead involved the beta decay of the element cobalt. The nuclei of cobalt atoms spin like a top and by cooling down to temperatures of near absolute zero, so that the nuclei do not wobble, and by placing them in a magnetic field, it is possible to make all of the nuclei spin in the same direction (let's call this anticlockwise). If the spin is thought of as like a screw, there is a hemisphere into which the screw would screw in and the other hemisphere would be that from which it unscrewed. If some of the cobalt nuclei decay radioactively, the beta particles (electrons) are emitted preferentially into the screwing "in" hemisphere (let's call this "upwards"). Why this is need not concern us, what we are interested in is how this would appear in a mirror. As Fig 10.1 illustrates, in the mirror the electrons would still be emitted upwards but the nuclei would be rotating in the opposite direction, clockwise. The question is whether this mirror image can also occur in the real world: if it does not then we have a way to distinguish the real world from its mirror image.

The first question is how to set up the mirror image of the experiment in reality. It turns out that if all the electric currents and other paraphernalia that go into making the magnetic field are reversed, so that the magnetic field itself is reversed, the cobalt nuclei will all spin in the opposite direction, as their mirror images did in the first experiment. So far everything has been changed in such a way that it looks like the mirror image of what we had before. Now we need to see what happens when the radioactive decay occurs: are the electrons emitted upwards or downwards? If everything is indeed mirror symmetric, we would expect the electrons to be emitted upwards. The surprise discovery was that in reality the electrons were emitted preferentially *down*wards and

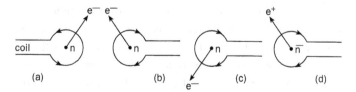

Fig. 10.1 Breaking the mirror: (a) electrons are emitted upwards; (b) its mirror image; (c) we see the result of the actual laboratory experiment where the electron is emitted in the opposite direction; (d) the matter–antimatter image of (a) which does occur in Nature, (see Chapter 11).

not upwards as mirror symmetry would have implied. Here for the first time is a way of making an absolute distinction between reality and mirror images.

Today we know that beta radioactivity emits not only an electron but also a ghostly little particle called a neutrino (Italian for "little neutral one"). Neutrinos (page 146) are so lightweight that if they have any mass at all it is so small as to be indeterminable; it is possible that they are massless. Beta radioactivity is an essential part of the energy production in the sun and neutrinos from the sun are streaming down on our heads by day and up through our beds by night, undimmed, as the earth is almost transparent to them. A massless neutrino, with no electrical charge to give its presence away and hardly interacting with anything as it rushes through space at the speed of light, sounds like as near to nothing as one can get. However, it has one bizarre property that is at the root of the mystery of the mirror universe; it spins like a top, similar to the cobalt nucleus, but with one important difference. It turns out that as the ghostly neutrino corkscrews through space, it spins only as a left-hand screw—there is no such thing as a right-handed neutrino in the universe, so far as we know. If you saw an image of a left-handed neutrino, you would be looking at something in the real world; a right-handed neutrino by contrast exists only in the mythical universe behind the mirror. The left-handed neutrino is as near as we have yet come to revealing the left-handed Creation at the level of the basic particles and the forces that form them into bulk matter.

Since 1956 evidence for this intrinsic mirror asymmetry has shown up in other phenomena, sometimes where neutrinos are involved and sometimes not, but the common feature is that in all cases the weak force is involved. It not only appears to be an intrinsic feature of nature but also to have been present since the earliest moments of the universe.

Although the sinister weak force is indeed "weak" under normal circumstances here on earth, it becomes more powerful as the temperature grows (page 151). In Chapter 8 we met the Z^0, "heavy light" (see page 154). At CERN, the LEP (Large Electron Positron) accelerator recreated the conditions 10^{-10} seconds after the universe was born, when the temperature was $10^{15°}$ and the Z^0 was acting in its full glory. These experiments have shown that the intrinsic asymmetry between left and right was present and significant then, having become enfeebled only as the universe cooled to its present state. The scientists then realized that as the Z^0 can carry a weak force between electrons and protons without need for neutrinos, even at room temperature, it will influence the constituents of atoms and molecules over and above the prominent electromagnetic force that binds them. The Z^0 subtly disturbs the motions of electrons in atoms but does so in a way that spoils mirror symmetry. Within atoms, the Z^0 disturbs electrons that are spiralling left-handed slightly more than those that are right-handed.

In recent years, physicists have carefully studied atoms at room temperature and managed to discern the trifling effects of the Z^0, swamped though they are by the dominant electromagnetic force. It transpires that there is indeed a subtle left-handed memory in processes that involve electrons and protons, as in the structure of atoms.

Now, electrons are ubiquitous in the atoms of our bodies and all matter, which raises the question of whether matter at large retains a significant "sinister" memory. Some scientists have suggested that this subtle asymmetry, though trifling, could, on the huge time-scales of evolution, have led to the left–right asymmetry in living things. This brings us naturally to the question of whether the chirality in organic molecules could have been imprinted during the cosmic era, within the first 10^{-10} seconds of the Creation.

Biologists, chemists, and atomic physicists have traditionally assumed that the electromagnetic force is the only fundamental force capable of producing chemical effects. Were this the case there would have been no intrinsic preference for right or left and the observed chiralities of amino acids in living things would have had to arise by mechanisms such as we have discussed already. Even though we now know that the Z^0 causes subtle differences in the make-up of certain molecules and their mirror images, its effects have been supposed to be so small at low temperatures that they can be ignored.

At first sight this indeed appears to be so. A chiral molecule and its mirror image (which, as explained in Chapter 4, are defined by the way they rotate the plane of polarized light) are almost equivalent in terms of stability. The most stable states tend to be those with the lowest energy. The effect of the Z^0 within atoms is such that it makes one of the chiral forms some 10^{-19} eV lower in energy, and hence with greater stability, than its mirror image. (An eV is a unit of energy; it is the amount of energy that an electron gains if accelerated by one volt.) That is the good news; problems arise as soon as one realizes how trifling this extra "stability" is. At room temperature the molecules of matter are moving around with kinetic energies of about $1/100$ eV each; this implies that as a fraction of the whole, the 10^{-19} eV difference in energies of the two mirror-image molecules is 10^{17} times smaller than their intrinsic thermal agitation of $1/100$ eV. Such fine effects are barely detectable in carefully controlled experiments on individual frozen atoms; the odds against them being overwhelmed by thermal agitation seem astronomical.

However, there are tantalizing hints that all may not be lost. Computations have been made for four amino acids which show that it is the left-handed versions that have the lower energies and hence would be expected to be the more stable. This appears also to be the case in nature, and so the question arises whether there is some way of amplifying this advantage over the course of time so that, for example, the 20 amino acids which make up the proteins could have converted almost entirely into the left-handed form.

The answer depends on what you believe to be important and what can be safely neglected when modelling the behaviour of molecules in the oceans and soup of the early earth. Some scientists have concluded that on long time-scales (of about tens of thousands of years) for reactions occurring in large volumes (such as the oceans), the effects of mirror asymmetry can prove to be significant, while others have questioned the assumptions made in these computations and argued they may have been oversimplified. However, Abdus Salam (who shared the 1979 Nobel Prize, having established the analogy between the weak and electromagnetic forces, in particular successfully predicting the existence of the Z^0) has pointed out that the phase transitions could amplify the effect.

Changes of phase are very familiar in everyday life and we have illustrated, in Chapter 9, what goes on in the familiar case of water freezing. At 0°C the water has less energy if its molecules are locked tightly together as ice crystals than when roaming free in the liquid state. As nature seeks the states of lowest energy, ice is favoured, and the excess energy is released as heat. The two phases, ice and liquid, can coexist at the "critical" temperature of 0°C.

At the critical temperature where phase changes are taking place, the distances over which the effects of a disturbance spread can be very large. This distance is known as the "correlation length." For example, in the case of a piece of iron cooling below the point at which magnetization can survive, if one of the constituent electrons is oriented such that its intrinsic magnetic "north" pole is pointing in a certain direction, then all the other spins within the same correlation length will align with it to a certain extent. There is an analogous correlation length in gas—liquid systems near their boiling points and this gives the size of the droplets of liquid that form spontaneously in the vapour phase or of the bubbles of gas that erupt in the liquid phase.

The "magic" of quantum mechanics causes phase changes to occur at low temperatures in some systems, leading to phenomena such as superconductivity or superfluidity. These occur when the atoms of the substance that had previously been moving at random suddenly "get into line," as if a well-drilled troop of soldiers on parade. Indeed, this analogy has been often used. Imagine that you are looking at a crowd of people from high above. They are all milling around; there is no order as each is doing something different. However, if instead the crowd were a parade of soldiers, then every one of them would be doing the same thing and it would be very much easier to see from a distance what it is. This is what can happen for a collection of atoms, where at a critical temperature large numbers can change from disorder to order remarkably rapidly.

Abdus Salam built a model based on this analogy, stressing that even though the difference in energy between the right- and left-handed states of the organic molecules was trifling compared to the energies of thermal agitation, the cooperation of large numbers of such molecules could cause the whole to "condense" into the lower-energy state at some critical temperature. He suggested that the molecules of amino acids could act collectively like the troop of soldiers, all rapidly falling into line in the left-handed configuration of lower energy. Whether or not this is how it really happened is still open to debate but that it was in principle possible is an interesting conclusion of Salam's work. In this theory, the coiling of our DNA is a fossil relic of an asymmetry that entered the universe within 10^{-10} seconds of its birth.

We are asymmetric. Chemists and biologists thrive on asymmetry as our amino acids and DNA are asymmetric and so it is surprising perhaps that physicists were astonished when the subatomic world proved to be asymmetric. Nonetheless, they quickly realized how a sense of symmetry could be restored, but to do so involved more than the matter that we are made of: the key ingredient was antimatter.

We will meet this bizarre stuff, beloved of science fiction, in the next chapter, but for the moment all that we need to know is that particle and antiparticle are identical in all respects but one: they have the same mass, the same size, and the same amount of electrical charge, but if the particle has negative charge (as for the electron) then the antiparticle has positive (hence the "positron"). We can make a metaphor for the intimate relation between electron and positron, matter and antimatter, by the imagery in Fig. 10.2: if the particle is represented by some shape, the antiparticle is likened to the negative image. The electron spins as it travels

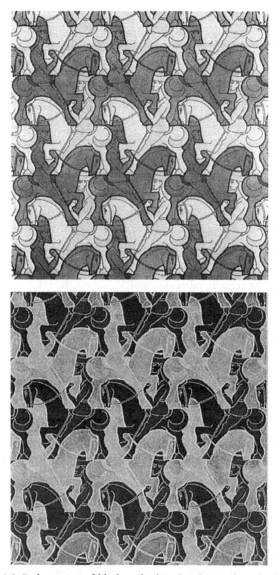

Fig. 10.2 Escher print of black and white knights on horsebacks (a), together with its negative image (b). View the latter (b) in a mirror and compare with the original (a). MC Escher's Symmetry Drawing E67 © 1999 Cordon Art BV, Baam, Holland. All rights reserved.

and if the image refers to an electron spinning as a left- (or right-handed) screw, then the positron in the corresponding circumstances would be spinning right- (or left-handed). In the case of the neutrino, which always spins like a left-handed screw (page 192), its antineutrino counterpart always spins right-handed. Thus it is that while the basic processes of radioactivity, in particular those involving neutrinos, have a clear "left-handed" behaviour, the corresponding processes in the antiworld, involving antineutrinos for example, are "right-handed." The picture had a certain symmetry to it: real world left-handed and antiworld, right-handed. We can use the metaphor of the negative images to illustrate this.

The replacement of a particle by its antiparticle is called "charge conjugation" symmetry (or simply "C" symmetry), while the conventional mirror symmetry (or "parity") is denoted by "P." Performing both operations is known as CP. If left-handed particles behave like right-handed antiparticles always, then nature would be "CP" symmetric. Where the weak force skews particles to a sinister left-handedness, it gives the corresponding antiparticles a right-handed dexterity; in a profound way the combination of both mirror and charge reflection balances the books. A way of visualizing this follows.

I have long enjoyed the works of Escher, the Dutch artist whose graphic drawings use realistic detail to achieve bizarre conceptual effects. One of his works fills the entire canvas with knights on horsebacks: white knights to the left while the space between them turns out to have exactly the same outline and, shaded black, is seen as black knights riding to the right (Fig. 10.2a). This picture has a certain symmetry but it is not mirror symmetry; in the mirror you would see white knights heading right and black knights to the left.

Think of the black and white knights respectively as portraying matter and antimatter: the picture then portrays the tantalizing relation between the two. If you took a negative image of this, replacing the black by white and vice versa, you would obtain Fig. 10.2b. Now look at this in a mirror and compare with the original in Fig. 10.2a: you will be back where you started. The image is not mirror symmetric ("parity" fails), nor is it "C" symmetric, but put them both together and you preserve the beautiful image; we say that the image is "CP" symmetric.

Thus it was that although matter skewed to the left, antimatter restored the symmetry. We are made of the asymmetric half of a perfect whole—at least that is what the scientists thought, briefly. Then, in 1964, an experiment at the Brookhaven National Laboratory in New York overthrew this paradigm. The ensuing Nobel Prize, given to James Cronin and Val Fitch, was described in one Swedish newspaper as being for their discovery that "The laws of nature are wrong"! We will come to this in Chapter 12—after we have met antimatter.

Chapter 11

Antimatter matters

Antimatter has an aura of mystery, the promise of a natural tweedledum to our tweedledee, where left is right, north is south, and time runs in reverse. Its most celebrated property is its pyrotechnic ability to destroy matter in a flash of light, converting the stuff that we are made of into pure energy. In science fiction, antiplanets tempt travellers to their doom even as antihydrogen powers the engines of astrocruisers.

Antiparticles are generally considered to be exotic, made only in experiments in high-energy accelerators and somehow "not of this world." Yet antimatter is real and its hand-in-glove mirroring of matter is the nearest thing we know to perfect symmetry. Indeed the first glimpse of the antiworld came not from experiment, a chance discovery, but from the beautiful patterns that the English mathematician Paul Dirac saw in his equations. As crotchets, minims, and semiquavers on a stave are mere symbols until interpreted by a maestro and transformed into sublime melody, so can arid equations miraculously reveal harmony in nature.

When Dirac entered the scene there were two great theories of the universe that seemed quite unrelated: quantum theory and relativity. What Dirac did was to marry them. It was thus that he stumbled onto the antiworld.

Quantum theory describes the world within the atom, the world of the basic particles such as electrons. Atoms are often visualized as miniature solar systems, with the planetary electrons whirling around the nuclear sun: little things whizzing around something big in the middle. The actual solar system involves the planets, attracted by gravity, while an atomic nucleus attracts the electrons by electrical forces. However, as soon as this picture was first proposed people worried about it. The electrons should have spiralled into the nucleus, so rapidly in fact that atoms could never have survived long enough for us to be here.

It was only after quantum theory appeared that the solution emerged. When you get down to distances smaller than a millionth of a millimetre where atoms live, experience as we know it on the large scale of everyday experience tends to be more subtle.

The quantum theory had been invented by Max Planck in 1900 but a quarter of the twentieth century had elapsed before the quantum rules describing the behaviour of electrons in atoms were discovered. The Austrian, Erwin Schrodinger and the German, Werner Heisenberg wrote down their equations that successfully described the behaviour of atoms and molecules through the choreographed dances of their constituent electrons. The consequences of the equations, the extraction of their full implications would be, and still are, major computational tasks but, at heart, it appeared as if little else needed to be understood.

While quantum theory describes the world of the particles, relativity describes what happens when things approach the speed of light. In the quantum world, common sense, as developed in our large-scale experiences, ceases to apply and particles appear less as things of substance and more like waves, with the disarming potential to be in more than one place at a time. Similarly, in the world that Einstein described with his relativity theory, objects become infinitely heavy as they approach light speed and again common sense is stretched.

In our everyday experience these phenomena do not seem to be right, yet all the evidence suggests that the universe does indeed behave in this way. (For an eloquent description of these marvels in everyday life read *The Quantum Universe* and *Einstein's Mirror*, both by Tony Hey and Patrick Walters.) The theories of relativity and quantum mechanics exist, like all theories, to be extended and tested until they are proved wanting. Fundamental physical science involves observing how the universe functions and trying to find regularities that can be encoded into laws. To test if these are right, we do experiments. We hope that the experiments won't always work out, because it is when our ideas fail that we extend our experience. The art of research is to ask the right questions and discover where your understanding breaks down. This is what Dirac was doing in 1928. He was trying to extend and test the existing theories of the universe and to see where this led. Specifically, Dirac asked himself what happened when one tries to deal with the simplest particle of all, the electron, using *both* quantum theory *and* relativity at the same time.

Paul Dirac's father was Swiss; he moved to Bristol and taught languages. French and English had equal place in the home and Paul was brought up bilingual, although he was unusually taciturn in both languages. Stories of his linguistic economy are legion, and at college dinners there was always the delicate issue of who

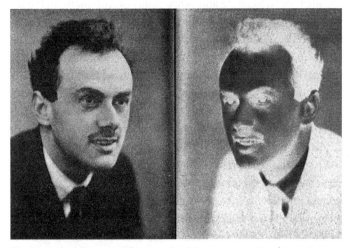

Fig. 11.1 Paul Dirac and his negative image representing his antimatter self. (Mary Evens Picture Library.)

would have the mixed privilege of seating by the silent genius. On one occasion, when E. M. Forster was a guest, the college had the inspiration of sitting the pair together; Forster also was easier with written words than conversation and Dirac was an avid reader of Forster's works. According to the folklore, which is probably apocryphal but could have been true given the characters, the evening developed as follows. Through the soup course nothing was said but as the main dish was served, Dirac leaned over and, in a reference to Forster's *Passage to India* asked "What happened in the cave?" This was to be Dirac's contribution to the evening. Forster pondered Dirac's question but sat silent. He ate on and ruminated further. Finally the dessert arrived and Forster delivered his answer: "I don't know."

Limited though Dirac was in personal communication, like Forster he expressed himself through written symbols. For Dirac these were the hieroglyphs of mathematics and, in the opinion of all theoretical physicists, the creativity, power, and elegance of Dirac's mathematical expressions compare with Shakespeare or Beethoven. In his great work of 1928, where he brought together the ideas of Heisenberg and Schrodinger on the quantum theory and fused them with the other magnum opus of the century, Einstein's relativity, Dirac invented a whole new mathematical language. Strange and bizarre it appeared at the time but today it is part of the education of students and used throughout theoretical physics. The result that convinced all of its essential truth was its explanation of the last of the atom's puzzles—the apparent property whereby electrons spin around their own axes like miniature tops.

Even if that had been the total of Dirac's insights it would still have been singular among theoretical physics in the mid-century, completing the description of the atom and leading to the (transient) speculation that physicists could now become biologists, as all the rules had at last been discovered. However, the greatest implication of Dirac's equation (as it will be known for all time) was that it opened a horizon to an entirely new world, one that does not exist around us here. Dirac's spinning electron, which nature had revealed to us through subtle fine structures in the spectra of light emitted by atoms in magnetic fields, cannot exist alone; his equation turned out to have two solutions, one being the familiar negatively charged electron, discovered in 1897 by J. J. Thomson (see page 104), while the other implied the existence of a bizarre mirror-image version, identical in all respects save that its electrical charge is positive rather than negative. This "antielectron" is more succinctly known as the "positron" (for "positive electron") and is an example of antimatter.

Dirac's prediction of the antielectron seemed to many at the time to be science fiction: up until then the only particles known or predicted existed as constituents of the matter around us, namely electrons in the periphery of atoms and protons and neutrons comprising the atomic nucleus. The antielectron had no place at all and at every seminar people would ask Dirac "where is the antielectron?" Such questions invoked laughter and Dirac soon tired of them. It was particularly galling that few of his contemporaries had seriously studied his equation let alone were able to follow his arguments. To deflect these persistent questions Dirac announced that as the proton has the opposite charge to the electron, it could be a candidate for being the antielectron.

From this has grown the folklore that Dirac seriously considered the proton as a possibility for the positively charged particle that had emerged from his equations like a rabbit from a magician's cloak. To modern physicists this seems absurd as the proton's mass is some 2,000 times greater than that of the electron and the mass of particle and antiparticle must be the same. The deep symmetries underlying matter and antimatter are much better understood today than in 1928 and it is moot whether the mismatch in masses was so obvious a nonsense as it now appears. Peter Kapitsa, the famous Russian physicist and contemporary of Dirac, claims that Dirac made the remark as a joke, his taciturn character notwithstanding, in order to quieten the persistent questioners and deflected the mass questions as a "detail" to be solved.

Whatever the reality, the questions ended by 1932 when the anti-electron was found in cosmic radiation, with positive charge and identical mass to its electron sibling. The massive nuclear proton is another beast entirely and, according to Dirac's equation, also has an "anti" counterpart: a negatively charged antiproton. The antiproton took longer to find, eventually being produced in intense collisions of protons and other nuclear particles at the high-energy accelerator in Berkeley, California, in 1955. Dirac's prediction that to every particle variety there exists an antiparticle counterpart, is now recognized to be an essential truth, a glimpse of a profound symmetry in the fundamental tapestry of the universe.

Positrons at work

Once Dirac had pointed the way, positrons turned up all over the place. The flash of energy resulting from their annihilation by electrons enables PET (positron emission tomography) scanners to trace the passage of chemicals through the brain. In industry, a similar process enables engineers to detect early signs of metal fatigue in jet engine turbines.

It was in 1932 that Carl Anderson, at Caltech, first reported his sighting of a positron in the cosmic radiation passing through his cloud chamber. On hearing the news, other physicists immediately looked through their old photographs from cloud chambers and found evidence for the positron in them that they had previously overlooked. Among those who had missed the positron were Irene and Frederic Joliot-Curie. Having missed one Nobel Prize with the neutron, the Joliot-Curies realized that they had missed the positron also. In 1936 Anderson was awarded the Nobel Prize for his discovery of the positron.

As beta radiation—the emission of electrons from atomic nuclei—was by then well measured (it was this essentially that Becquerel had stumbled upon in 1896), it was natural to suppose that positrons also might be produced in some radioactive decays.

Here the Joliot-Curies were successful, in 1934 verifying that certain nuclei can indeed spontaneously emit positrons.

The major practical difference between radioactive decays that emit positrons and those that emit electrons lies in what happens next. An ejected electron may flow as current or join in the dance of planetary electrons in neighbouring atoms, later to initiate chemical reactions and countless other adventures in the future of the universe. A positron, by contrast, is a stranger in our land and isn't long for this world. It finds itself surrounded by hordes of negatively charged electrons. Momentarily one of these electrons partners the positron in a cosmic dance of doom as they encircle one another and, within a microsecond, mutually annihilate in a flash of light. It is this that is the key to the practical application of positrons.

Positron emission is natural and common; it is the ability of certain nuclear isotopes to emit them that is so useful in medicine and technology. Some examples are carbon-11, nitrogen-13, and oxygen-15 which are radioactive forms of common elements in the body and can be used, along with the positron emission, to trace bodily functions such as in the brain. The basic principle is that when the nucleus emits a positron and the latter annihilates with a nearby electron, two gamma rays can emerge almost back to back. This pair can be detected using electronic circuitry developed in particle physics and thereby one can locate the emitting nucleus very accurately. Now for the applications.

When you are thinking, various parts of the brain are active to different degrees. Being active uses energy which is supplied to the brain as chemical sugars in the bloodstream. If we could measure the concentration of sugar within the brain it would give some indication of the brain's activity. Chemists can incorporate radioactive atoms into sugar molecules and those sugars can be ingested and distributed within the body to the regions that are active such as the heart, lungs, muscles, and brain. The essential idea that has revolutionized aspects of medical diagnosis is to use

Fig. 11.2 Trails left in a particle detector by an electron and positron. They spiral in opposite directions, hinting at their subtle symmetry. (For more images of positrons see *The Particle Explosion* by Close, Marten, and Sutton published by Oxford University Press.)

sugars that radiate positrons. The positrons are immediately annihilated by ubiquitous electrons in the atoms in the vicinity. The debris from the electron—positron annihilation consists of two gamma rays—flashes of high-energy light—and we can tell where in space the annihilation took place, and hence where sugar was located, simply by looking at the gamma rays that come flying out.

By surrounding a patient's head with a halo of cameras one can build up images of the brain in slices—the technique known as positron emission tomography or PET. The particular isotopes of interest tend to be rather short-lived (for example, oxygen-15, used to label oxygen gas for study of oxygen metabolism, has a half-life of only two minutes) and so they need to be prepared near to the patient. They can be made by bombarding suitable elements with beams of protons from an accelerator. Small accelerators used for nuclear research are made available for medical research or hospitals may have a small, dedicated accelerator for the purpose.

Positron annihilation is also useful for studying materials. One example involves annihilation in metals that can reveal the onset of metal fatigue in advance of other techniques. This has been used in testing aircraft turbine blades, enabling safety margins to be narrowed and profits increased.

So antiparticles, in the form of positrons, are familiar and put to use daily. They are less familiar than electrons merely because they are so outnumbered that they are rapidly killed off. They are also produced at the centre of the sun in the first stage of its fuel process. The sun consists primarily of hydrogen plasma, protons and electrons roaming independently as electrically charged gases. When two protons bump into one another their high energy can convert one of them into a neutron and a positron. The neutron and proton bind into a tight unit known as a deuteron, the nucleus of "heavy hydrogen" or "deuterium"; two deuterons subsequently colliding forms a single nucleus of helium which is the major end product or ash of the solar fusion process. The positron is almost instantly annihilated by the electron plasma and gamma rays are produced. It takes a hundred thousand years on average before these gamma rays have made their way, buffered by the plasma, to the surface; by this time they have lost energy and are radiated as visible light that we see as the glowing ball in the sky. Thus daylight is an end product of positrons.

Positrons are relatively easy to make. When electrons pass through the electric field of a heavy atomic nucleus, the field is disturbed and can materialize pairs of electrons and positrons. This is simple enough to arrange and is the step in a promethean quest at CERN in Geneva where scientists and engineers recreated the conditions of the aftermath of Creation. If the newly created electrons and positrons pass through a magnetic field, the negatively charged electrons and their positively charged positrons are deflected in opposite directions. Thus one had two beams, one of electrons and the other of positrons. To keep the positrons alive in what is for them a hostile environment of matter involves containing them in a tube where the vacuum is even better than on the moon. (This is the first of many technological near miracles that have to be overcome if we are to journey to the start of time.) The beams are then accelerated by electric forces and steered by powerful magnets around the circular tunnel known as LEP (the "Large Electron Positron" collider); at near the speed of light the two beams are smashed into one another. The head-on collision between these elementary pieces of matter and antimatter creates a momentous annihilation. The energy in the tiny volume where this happens, at this unique laboratory on earth, is similar to what was present everywhere throughout the entire universe in the first moments after its birth.

By recreating the conditions of the start of the universe in this way it is possible to study how things were at an epoch when matter was first being created. Everything about these experiments shows that particles of matter and of antimatter emerged in equal amounts, perfectly balanced. Here, where relativity and quantum theory met, we see revealed their legacy: the perfect symmetrical emergence of the primeval particles of matter and antimatter.

The symmetry of the Creation

At the end of the 20th century LEP was the world's largest scientific instrument; 50 metres below the surface, in a tunnel that was as long as the Circle Line on the London Underground, magnets steered electrons and positrons to their goal. I visited LEP at the same time as a group of students from Britain. We had descended in a special elevator down a shaft that was as tall as a skyscraper—though in this case going the equivalent of 20 storeys deep into the ground. We arrived at the ring of magnets, which curved gracefully until they disappeared around the bend into the remote distance of their 27-kilometre journey.

We could have walked all the way around the circuit if we had had the time. Bicycles enabled the technicians to reach distant equipment faster and a cable car suspended from the roof of an access tunnel enabled materials to be transported at a steady 20 miles per hour. Even with these aids to help you, you needed to allow an hour for the whole journey. Compare this with the particles in LEP which were travelling at 99.9999 per cent of the speed of light: they made 11,000 circuits every second. The electrons sped beneath Swiss vineyards, crossed the international border into France, scurried under the statue of Voltaire in the French suburb of Geneva where he spent his final years, rushed beneath fields, forests, and villages in the foothills of the Jura mountains, until they smashed head on into their antiparticle opposites, the positrons. There were four places around the circle where these collisions occured and special instruments recorded the results. At each of these sites the tunnel suddenly opened into an underground cavern with an internal volume that was bigger than Notre Dame.

That was the scene awaiting us as we followed the magnets around the bend in the tunnel. Into view came a wall of concrete that appeared to signal the end of the line. The ring of magnets stopped; a small tube that contained the beams emerged from within the magnets and disappeared through a small hole in the

Fig. 11.3 LEP ring superimposed on the Geneva landscape. (Courtesy of CERN.)

centre of the concrete wall. To the right there was a door and as the first student passed through I heard an exclamation, "wow!," as the vast chamber that housed the detector came into view.

There were eight chambers like this around the LEP ring of which four contained detectors, one at each of the four collision points. Each was like a huge and highly complex piece of jewelry, full of crystals; they had poetic names that recorded the glory of their task—Delphi, Opal, Aleph, and, more prosaically, L3. They were described as the ugliest, hairiest, most complicated big machines ever made. (One visitor, staring up at the titanic dimensions of the multi-storey Delphi, with its metal, industrial catwalks and stairways ascending four storeys, described the sensation as similar to what a good-sized weevil must feel like when confronted by a loaf of bread.) Each detector had some reason to be in the *Guinness Book of Records*: for example, Delphi had the biggest superconducting magnets in the world, while L3 had an electromagnet with more iron than the Eiffel Tower.

Each one was, in essence, a 10-metre-long cylinder consisting of state-of-the-art electronics. Myriad flashing lights that warned of high voltages of intense magnetic fields greeted you as you climbed to the top of 200 metal stairs. There were thousands of cables stretched beneath your feet, and wires cascaded down the

Fig. 11.4 Inside the LEP tunnel. (Courtesy of CERN.)

sides of the detector giving it the appearance of wearing metal dreadlocks. You could then walk along a gantry that allowed you to stand atop the detector, at its centre point, five metres above the point where the positrons meet their nemesis.

Until the LHC began this was the nearest that humans had ever come to the Big Bang. The two beams had to be in the right place at the right time for this to work. The electrons and positrons had entered along the central axis of the cylinder, the positrons from one side and the electrons from the other, and when they met they annihilated in a flash of energy. That was the key moment; for a fraction of a second you were replaying the Big Bang at the start of the universe. Out of that flash of energy streamed particles of matter and antimatter, just as they did originally at the start of time. The detector encircling the spot captured and recorded the first moments as these primeval pieces of matter emerged, repeating over and over the long-ago act of the Creation.

As the particles streamed across the face of the detector, they triggered the electronics, the ensuing signals instantly passed along cables into the computer room nearby. The magnets in the detectors were so powerful that they distorted the images on computer screens several tens of metres away. On the first day of operation no one anticipated this; to the astonishment of the scientists watching, the images on computer terminals near to the detector were turned upside down and spiralled around by the intense magnetic fields.

The screens were like eyepieces at the end of a telescope; they recorded and displayed the trails of the newborn particles and antiparticles and the scientists then had to decode them, interpreting these strange hieroglyphs from their experiences in this bizarre world. In effect it was like trying to reconstruct the scene of a car smash from the skid marks on the road.

Hundreds of scientists from all over the world collaborated in these experiments; they needed the data instantly and were able to communicate the results of their analyses to one another easily. The World Wide Web was invented at CERN in order to do this. A revolution at its creation in 1989, today it is the "printing press" of the twenty-first century. It is when the images were displayed on those screens that the scientists were able to work out what went on at the scene of the smash. This was where the computer revolutionized particle physics. Thirty to forty years ago you might have looked through maybe 500 pictures a year, searching for the one or two that were really critical; miss them and your Nobel Prize was lost. Nowadays, the computer can analyze thousands of these images every hour and furthermore can be programmed to search and recognize the really interesting ones.

Fig. 11.5 A LEP detector with four scientists setting the scale. (Courtesy of CERN.)

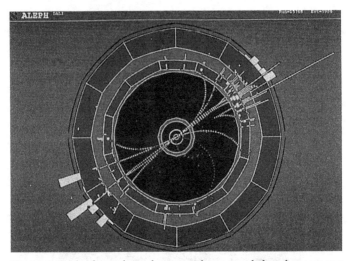

Fig. 11.6 Trails of particles and antiparticles as revealed on the computer screen; compare the computer view with the end view of the detector in Fig. 11.5. (Courtesy of CERN.)

Designing and building the detectors was a feat of engineering that was as remarkable as anything that the experiments discovered. Nothing in the detectors existed in the electronic market-places of the 1980s when they were first designed. The entire colossus was designed by hundreds of people in teams around the world; the individual components, down to nuts and bolts, were made in different places, even in different continents.

LEP and its special cameras were so precisely engineered that they could detect the motion of the moon. Early on the scientists discovered that the electrons and positrons arrived fractionally late and at other times fractionally early for their assignment of mutual destruction. The mismatch was less than a nanosecond—that is one thousandth of a millionth of a second—but nonetheless the precision of LEP was sensitive to it; the timing went from early to late and back again on a 28-day cycle. Then the scientists realized what was happening. The monthly lunar cycle raises tides in the oceans that can be several metres in height, but it also affects the rocks in the earth's surface, though by trifling amounts. The 27-kilometre length of LEP was expanding and contracting by a few millimetres every month, hence the beams had a little further to travel at some times, and a shorter circle two weeks later.

They recalibrated the computer software to take account of the monthly tides and settled back to what they now anticipated would be a regular constant behaviour. This was when a second anomaly showed up, one that had previously been hidden by the lunar effect. The technicians controlling LEP noticed that it seemed to go in and out of tune for a few minutes each morning and again in the evening. This was a real mystery, until there was a national strike in France, at which point the anomaly disappeared; once the strike was over the anomaly returned. Then someone realized that the timing coincided with the arrival of the French high-speed train, the TGV, in Geneva; as it draws power from the overhead wires the sensitive electronics at LEP were reacting to it. Having corrected both for the moon and the TGV, the machine ran in perfect tune for months each year, limited only by shut-downs for upkeep and occasional alarms, as when a disgruntled worker rammed two beer bottles into the accelerator tube during a magnet overhaul.

During the years of LEP's operation more than 10 million pictures were taken showing the Creation of matter as it was within a billionth of a second after the original Big Bang. Gradually we began to decipher their story.

So what was seen in the aftermath of the intense annihilation of LEP's electrons and positrons? Sometimes two gamma rays emerged, as in the PET scanner. However, in the intensity of the Big Bang, it turned out that there were many more things going on and LEP uncovered them. Out of the "mini-Bang" emerged particles and antiparticles (such as electron and positron) or quarks (the seeds of atomic nuclei) and their antiparticles (the "antiquarks").

Some of these were known about before LEP was built but what LEP allowed us to do was to clarify how all the particles related to one another. First, there were the basic seeds of earthly matter: the electrons of our atoms and the "up-and-down" varieties of quarks (see page 150) that formed its nuclear seed. The way they showed up in the detectors at LEP and behaved as they mingled, indicated that they were very similar in their fundamental characters. It is these experiments that showed the nature of the early universe to have been more uniform, more symmetric, than is the case today. I have summarized this already in Chapter 8 and to recap and reinforce the message, it is that nature has hidden its basic symmetries by burying the quarks first inside the protons and neutrons, then in the nucleus at the heart of atoms which link together as molecules and finally as bulk matter.

More than 10 years of experiments at LEP confirmed the similarity between the electron, neutrino, and up-and-down quarks time and again. But it showed more than this. It turns out that nature is not satisfied with the electron alone; there are two heavier versions known as the muon (some 200 times heavier) and the tau (nearly 4,000 times heavier) which are identical in electrical charge and all else but for their masses. There are three varieties of neutrino too, each of which seems to be a sibling of the electron, muon, and tau respectively. Although the non-zero masses of the neutrinos are too small to quantify, there are hints that here, as for their electrically charged cousins, it is the different masses that distinguish them.

This tripling also occurs for the quarks. The down quark has two heavier forms, called strange and bottom (the latter sometimes known more poetically as "beauty"); the up quark having its two heavier analogues named charm and top (the "beauty" school of nomenclature referring to the top as "truth"). The top (or truth if you prefer) weighs as much as an atom of gold and was too heavy to be created at LEP. (The Fermi Laboratory near Chicago has an accelerator of protons with enough punch to make top quarks, and it was there in 1994 that this ultimate concentration of matter was discovered.) Millions of examples of the others were made and studied at LEP and the message appears to be that they are identical in all essential respects but for their masses.

Why nature chose three sets is a mystery. How nature concentrates the great mass of the top quark into a volume smaller than we can measure is another of the great unknowns. It was hoped the answers might emerge when LEP was replaced by the Large Hadron Collider (LHC). Instead of mere electrons and positrons the LHC uses protons that split apart in more dramatic fashion and with greater energies, getting deeper towards the start of time (see Chapter 13).

The message from our search for the ultimate symmetry of the Creation is as follows. In LEP, where we set up the conditions of the original moment of the universe when matter and antimatter first appeared, we find that the electrons and the quarks seem to behave the same, as if there is, deep down, only one type of matter underwriting it all. This hints that some more profound symmetry, a "grand unified" description of matter and forces, was present early in the Creation and has become lost sometime soon after the original Big Bang, There is, though, one great mystery that became deepened by LEP's visions. The deep symmetry between matter and antimatter, embodied in Dirac's equation and verified in decades of experiments, is in stark contrast to the structure of the known universe now which consists of matter to the exclusion of antimatter.

The contrast between this asymmetry in the present universe and the symmetry that we associate with the Big Bang where matter and antimatter play equivalent roles, is striking. As if by some fundamental Darwinian selection we are made of matter, as are the heavens and all that we see, and would be annihilated were we to encounter our opposite in the form of bulk antimatter. That the universe is lopsided appears to be critical for our existence. It is also perhaps the greatest of the outstanding mysteries.

Chapter 12

Back to the future

Years ago at school, in the days before video, there was a film club where we would watch "real" movies. Someone, for a dare, had unwound one of the reels and then rewound it in reverse, creating much mirth as we watched a dead cowboy spontaneously come to life and leap backwards up onto his horse which then ran off hind first. It is usually easy to tell when a film is being played in reverse: cowboys "undie"; people grow younger; wrinkles turn to firm, youthful flesh; broken eggs spontaneously reassemble. These are all contrary to experience. The natural sequence is a process of order turning to disorder, of ageing, and decay.

It is easy to understand why this is when we realize that things are made of vast numbers of atoms. There is effectively an infinite number of ways those atoms could at random have been arranged, but only one that corresponds to the highly ordered object in reality. When an egg breaks, there are many ways that the shapes and sizes of the bits could fall but only one way that these bits could collect to form a whole, unblemished egg. Left to their own devices, the bits and pieces of matter are more likely to become disordered simply because there are more options available; it is this sequence of events, going from order to disorder that gives us our perception of the natural passage of time.

A simple and direct example is to watch a film of the start of a game of snooker. The 10 red balls are collected into an ordered triangle; the cue ball then strikes the pack and disturbs them.

Immediately there are innumerable possibilities for their recoil that makes every game of snooker unique after this opening shot. It is of course possible, though exceedingly unlikely, that the cue ball will leave the pack undisturbed and return precisely to its original spot. Were this to happen you could not tell whether you were watching the film of the real event or its time reversal; in general it would be immediately obvious as randomly moving balls tend not to gather themselves spontaneously into an equilateral triangle.

The natural direction of events is apparent with the 10 snooker balls, and with the vast numbers of atoms that make up macroscopic matter it is even more so. For individual elementary particles within atoms it is less obvious which way time is flowing. To see this in a familiar context, we can return to the snooker game when only the final black ball and the white cue ball are on the table. It is possible to play a "stun" shot where the white cue ball strikes the black, stops, and transfers all its momentum to the black. A film of this played in reverse would show the black ball hitting a stationary white one, coming to rest as the white ball recoils. Were it not for the fact that in real snooker it is the white ball that is used as the cue ball, one would be unable to tell if this was the real or a time-reversed event. Even then you could not be sure that it was not a real film of "negative" snooker where the cue ball was black and there were complementary colours for all the rest.

The arrow of time for bulk matter is a case of order becoming disorder overall without any change taking place in the individual elementary objects that make up the whole. The snooker balls appeared unchanged individually, it was their collective whole that gave away the reversal of the film. The flow of time from Big Bang, to life, death, and the eventual fate of the universe itself may be in a sense merely an illusion, the lottery of chance in the terpsichorean formations of countless atoms.

When one watches the individual components, as in the one-on-one snooker ball example, it is not so obvious—or may even be impossible—to tell which way the film is running. The received wisdom was that if one could reverse the direction of time in an interaction between two subatomic particles, the outcome would be a physically realizable process. Whereas the arrow of time is an illusion in the macroscopic world, at the most fundamental layer time appears to be a symmetry; the physics governing many natural phenomena, such as the interactions between elementary particles, permit time to run either forward or backward. The individual black and white snooker balls can collide white on black or black on white—either is allowed: the time-reversed film can be realized in the "real" world.

However, for 34 years there had been suspicions that time might not be perfectly symmetrical and, in 1998, physicists working at CERN in Geneva proved that this was the case. Although the atomic elements may care naught for the arrow time, there are strange particles, transient invaders in cosmic rays, that can tell left from right, and future from past, in an absolute way.

The tantalizing possibility is that this arcane discovery could indeed have implications for the existence not just of life, but for the entire material universe. The questions that we have been concerned with so far, such as why life cares about left and right, pale against the greatest conundrum of all: why is there something rather than nothing? The problem is this: theories and observation all suggest that in the first moments of the universe particles of matter and antimatter were created in equal abundance. Yet today the universe is not like this as everything that we know in the large-scale cosmos consists of matter. How this asymmetry came about is one of the great puzzles of both particle physics and cosmology. It is possible that time's slippery and enigmatic nature may explain why there was anything remaining after the Big Bang, why there is matter in the universe rather than a brightness of light and energy.

What's the matter with antimatter?

The established wisdom is that the universe began in a hot Big Bang some 15 thousand million years ago. Out of this energetic fire-ball, matter emerged along with antimatter. Today antimatter may be the stuff of science fiction but to physicists it poses a serious question: why isn't there more of it around? This perfect complementarity between particle and corresponding antiparticle, that is at the root of their mutual appearance from the initial fire-ball, also causes them to annihilate one another should they come into subsequent collision. In the dense cauldron of the infant universe such collisions would have been very common and this leaves a paradox: it is all very well to argue that in the first moment matter and antimatter emerged equally from the Big Bang, but an instant later they had annihilated one another. It seems strange that there is anything left at all.

The elementary pieces of an antiuniverse certainly exist: the positron was discovered in 1932, the antiproton in 1955, and the first antinucleus (the antideuteron, which consists of a single antiproton and antineutron bound tightly together) in 1965. These are produced transiently when energy is converted into particles and antiparticles following collisions between cosmic rays and the atmosphere, or between protons and nuclei in laboratories such as CERN.

Each of these antiparticles appears to be a precise mirror of its particle counterpart, with identical mass but with its electrical charge reversed. The ways that protons and antiprotons respond to electric and magnetic forces show how accurate this symmetry is in reality. Measuring their "weight" directly is not easy as the force of gravity is so feeble when acting on a single particle that it is at the limits of ability to measure it at all. However, it is relatively easy to compare the masses of two moving, electrically charged particles from the way their paths spiral in a magnetic field. Competing with the magnetic forces that are trying to steer the particle round a curve is the inertia of the particle that tries to keep it on the straight and narrow. This is similar to our experience as we attempt to steer a fast, heavy vehicle around a corner; the greater the mass the harder it is. We can compare the times for a proton or an antiproton to complete a circle in the magnetic field to get a precise measure of their symmetry. In the 40 years since the antiproton was first discovered, the precision of these measurements has improved to the point that today we know that the antiproton and proton are alike to better than one part in a billion.

Scientists will continue to improve measurements; some day perhaps this will expose patterns that are currently beyond our imagination. At present however the situation is that protons and antiprotons are precise mirror counterparts of one another to the best accuracy that we can measure. The same is true of electron and positron. All of this is as Dirac predicted—particles of matter and antimatter are perfect counterparts.

This knowledge implies astonishing consequences for the nature of atoms and matter in general. The simplest atom is that of hydrogen, consisting of a single, negatively charged electron encircling a single, central, positively charged proton. The electrical charges of these oppositely charged particles counterbalance so precisely that matter in bulk is, overall, electrically uncharged; it is gravity not electrical attractions that rules the motions of the heavens. Now imagine if all electrons were replaced by their antiparticle twin, the positron, while all protons were replaced by their antiproton counterpart. The overall electrical charge would still be zero and the attraction of opposites would still make a stable atom whose positron would encircle remotely the bulky, central antiproton just as, in the normal state of affairs, electrons encircle their central protons. This atom of "antihydrogen" is identical in its bulk properties to a hydrogen atom; only its internal make-up is switched.

Antihydrogen atoms are the most detailed "anti" matter clusters yet seen, though the theory and all experience implies that all atomic elements can exist in "anti" form, and that antitadpoles, antiplanets, and antimatter in all its forms are as realizable as the more familiar matter which dominates the known universe. Are antigalaxies of antistars surrounded by antiplanets of antimatter awaiting unsuspecting astronauts in the far reaches of the universe? How sure are we that there is no antimatter at bulk in the universe?

There are certainly no antimatter mines on earth except the likes of the vacuum tubes of LEP at CERN or the accelerator of

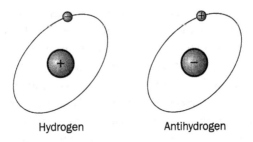

Fig. 12.1 Atoms of hydrogen and antihydrogen.

antiprotons at Fermilab near Chicago. These machines have been operating for around 20 years but after all that time no more than a milligram of antimatter has been made, used, and destroyed. It would take hundreds of thousands of years to make enough to drive an astrocruiser or to make diabolical weapons—even assuming you could store the stuff. The idea of bulk antimatter power is going to remain science fiction.

It is one thing to admit that there are no antimatter mines on earth, quite another to extrapolate from this that the entire material universe is made of matter to the exclusion of antimatter in bulk. How can we know the make-up of a distant star, seen only as a faint candle across the vastness of space? Its light comes from nuclear reactions within. In our sun these involve protons fusing to make helium and radiating light. In a distant antistar, antiprotons would fuse to form antihelium. Theory suggests that the spectra of the elements are the same as those antielements, and so by looking out into the night sky from earth we cannot tell stars from antistars. If in the future astronauts manage to travel to remote star systems, how can they ensure that they do not inadvertently land on an antiplanet and risk instant annihilation? What do we know about antimatter in the universe?

We know that the moon is made of the same stuff as astronauts whose joyful return proved that there is no antimatter up there. The solar wind, a continuous breeze of subatomic particles from the sun, hits the moon all the time. If the sun were made of anti-matter, we would be bathed in gamma rays from the moon due to the annihilation of matter and antimatter as this stream of anti-particles from the sun hit the lunar surface. The absence of such gamma radiation shows that the solar wind, and by inference the sun, is made of matter like us. Having established that the solar wind is made of regular particles we can see what happens when it hits the other planets in the solar system. There are no gamma ray flashes from these either and so we infer that the planets are also made from matter to the exclusion of antimatter.

We may not (yet) be able to reach the stars but we can know what they're made of because nature sends pieces of them to us. We are being invaded by extraterrestrials. News of matter, or antimat-ter, arrives from outer space in cosmic rays. If they pass through supersaturated vapour, like the air on a damp, cold, autumn morn-ing, a trail of drops form as they go. This cloud chamber, when surrounded by magnetic fields that bend the trails of electrically charged rays, can reveal the sign of their electric charge. This was how, in 1932, Carl Anderson first saw the trail left by an "electron" that curved the wrong way, thereby discovering the positron—the positively charged analogue of the electron and first example of an antiparticle—so vindicating Dirac.

Antiprotons and antineutrons have also been seen in cosmic rays. However, these are not remnants of antistars that have by mischance hit upon us but appear to be the debris formed from the energy released when a high-energy primary cosmic ray from outer space hits gas in the upper atmosphere. No antinuclei such as antihelium or anticarbon have ever been seen.

This latter point is particularly significant. We know that stars produce helium and so antistars would produce antihelium. Had an antistar exploded long ago and permeated the cosmos with antielements then there could have been antiplanets and antialiens breathing antioxygen, and antielements would arrive among the cosmic rays. Yet no antielements have arrived. The absence of any-thing as simple even as antihelium, by contrast with the abundance of individual antiparticles, suggests that these antiparticles have been produced in chance collisions between primary cosmic rays and the atmosphere; they are not the remnants of extraterrestrial bulk antimatter.

There are no signs of bulk antimatter. Through the most powerful telescopes we can see millions of galaxies distributed throughout the heavens. Some of these have close encounters and are distended as the tidal forces of gravity tug on their individual stars. If any of these colliding galaxies were made of antistars there would be distinctive gamma ray bursts at the boundary, but none have been seen.

It seems that we inhabit a volume of matter that is at least 120 million light years in diameter. It is tempting to extrapolate from this huge volume to conclude that the entire universe is made from matter. However, the universe is about 14 billion years old; on this scale our 120 million light years of matter is less than a tenth in radius, or one part in a thousand in volume of the entire cosmos. There is still lots of room out there for antimatter to be in remote corners of the heavens; it is not beyond the realms of possibility that an antimatter universe exists beyond the reach of our observations. So the puzzle of the missing antimatter is one of two: either we have to answer why the whole universe is lopsided in the sense that it is made of matter to the total exclusion of antimatter or we have to understand why it has settled into large and widely separated domains of matter and antimatter.

The idea that the universe is clustered into different domains (where we happen to live in one of the domains of matter) calls to mind the phenomenon found in magnets where a single piece of magnetized metal can consist of separate regions, or domains, where the directions of north and south poles differ, or more prosaically where diners at circular tables choose their napkins in a mismatched way (see page 166). If the universe consists of juxtaposed regions of matter and antimatter, the phenomenon of magnetism may provide an essential clue as to how this came to be. If the early hot universe consisted of an unruly mix of matter and antimatter and then suddenly cooled, could domains of matter and antimatter have emerged like the domains of magnetism as iron cools below its critical temperature?

When matter and antimatter meet they destroy each other. By the natural rules of chance there will be some regions where there is a slight excess of matter and other regions where there is a slight excess of antimatter. After the cataclysmic annihilations are complete, these regions will be separated domains of matter and antimatter. Whether these domains are large enough and sufficiently separated that such a structure can remain stable at its boundaries for billions of years is an unresolved question. The balance of the evidence however goes against the idea of a universe with vast regions of antimatter hidden deep in space. The general opinion is that antimatter was destroyed everywhere by some cosmic Darwinism in the first moments. "How?" and "why?": these are the questions.

In 1966 Andrei Sakharov realized that three conditions are needed for such an imbalance between matter and antimatter to arise. First, protons must decay, but so slowly that in the entire history of the earth the totality of decays would amount to no more than a few specks. The second involves the way that the universe cooled following the Big Bang and the third is that there must be a measurable difference between matter and antimatter.

So what is this difference? In Chapter 10 we illustrated CP symmetry by the analogy of Escher's black and white horsemen. Sakharov had come to this realization following the discovery two years earlier, by physicists in the USA, of a subtle difference that showed that CP is not an exact symmetry of nature. Continuing the Escher analogy, it was as if instead of the horsemen one had birds (Fig. 12.2). A negative mirror image is different from the original as the tails pointing up or down reveals. The symmetry between white and black geese, between matter and antimatter, is not 100 per cent exact.

To set the scene for this part of the story, let me introduce the main character, a short-lived strange particle known as the "kaon."

While the proton and neutron of atomic nuclei are built from quarks, and as such are pieces of matter, there are entities that consist of a single quark in a dance of death with an antiquark and whose relation to matter and antimatter is more subtle. Such combinations are short-lived because their quark and antiquark constituents rapidly come into contact and destroy one another. One of these is a beast known as a kaon whose mass is about half that of a proton. When it dies, the resulting debris consists of lighter particles known as pions which themselves eventually die out leaving more stable particles such as electrons and photons.

Even though these kaons live for less than a millionth of a second, modern electronics can trace them during their brief moment. It is in doing so that the failure of CP symmetry was first seen in 1964. Although kaons are not simply matter or antimatter, in that they contain both a quark and an antiquark, their inner

(a)

Fig. 12.2 Escher print of black and white geese (compare with Fig. 10.2). The negative image (b) viewed in a mirror (c) differs from the original in (a). MC Escher's Symmetry Drawing E18 © 1999 Cordon Art BV, Baam, Holland. All rights reserved.

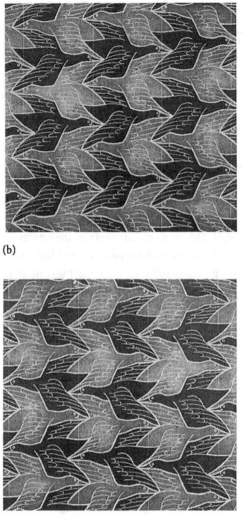

(b)

(c)

Fig. 12.2 (*continued*)

structure is unbalanced between matter and antimatter and it is this that gives them their special interest.

A kaon is built from one of the conventional matter quarks (the down quark) joined to an antiquark that is not of this world (a "strange" antiquark). This combination can be thought of as "matter-biased." The antikaon is the exact reverse of this. It has a conventional antimatter piece (namely a down antiquark) cohabiting with a strange quark (Fig. 12.3).

If both the C and P operations take place, a kaon will be turned into an antikaon and vice versa. So if you have a 50:50 mixture of kaon and antikaon (which may sound bizarre, but in the quantum world such a split personality is quite normal), and if CP is a symmetry of nature, then after the CP operation this combination will be unchanged. This turns out to constrain the way in which such a mix will behave, in particular how long it will live and what will be left behind when its brief life ends. One of the consequences is that

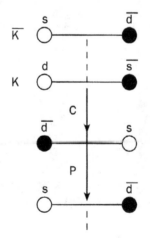

Fig. 12.3 A kaon and antikaon of d(down) or s(strange) quarks and antiquarks (the antiquark is denoted by a line above the respective symbol). The effect of C (quark turned into its "negative image," antiquark) and P (mirror reversal) is shown. A kaon has in effect been turned into an antikaon. This is analogous to the horsemen in Chapter 10.

these ephemeral beasts have two alternative fates. Either they die almost at once (in about one-tenth of a thousandth of a millionth of a second) or they have an elongated life of some five hundredths of a millionth of a second. These are known as the "short" and "long" lived modes respectively.

If CP symmetry is a reality then what remains after their brief lives can be an even or an odd number of pions. Specifically, the short-lived kaon will produce two pions while the long-lived ones give three. And that is how it is: almost. To understand better what is going on with these two alternative lifestyles, I will make a short diversion. If you are merely interested in the outcome, jump on two paragraphs.

In the illustration of CP symmetry with Escher's horsemen, it was sufficient to do the C and P switch just once to restore the image (Fig. 10.2). For a 50:50 mix of kaon and antikaon also, once is enough. However, for a kaon alone you would need to do it twice since the first time will have changed the kaon into antikaon, and you must then do it again, on the antikaon, to get back to where you started. Mathematicians would represent the sequence of actions by CP^2 and the action of starting with a kaon and ending with a kaon as equivalent to having multiplied your original kaon by 1 (since one times "anything" leaves the "anything" unchanged). Now, and here is the crunch, if $CP^2 = 1$, then the effect of CP done once can be represented by either +1 or −1 (since either of these multiplied by itself gives +1).

This piece of simple arithmetic is what turns out to be at work in the profound mysteries of quantum mechanics: a kaon and antikaon, in their 50:50 mix, can metaphorically carry either the +1 or the −1 label on their baggage. It is this quantity that is preserved when they die. It happens that a single pion, or any odd number of pions, carry always the label −1, whereas even numbers carry +1. If CP is indeed a true symmetry of nature, this number will be retained forever such that even or odd numbers of pions are formed when a kaon with CP of +1 or −1 dies. A kaon cannot turn into a single pion alone (energy would not be conserved), so three would be the minimum number needed for the CP = −1 mode. Making three pions is harder than two and so the CP = −1 mode resists decay longer than the CP = +1: thus the story of the short- and long-lived kaons. Now back to the main story.

In 1964, James Christenson, James Cronin, Val Fitch, and Rene Turlay, working at the Brookhaven National Laboratory on Long Island, New York, discovered to their astonishment that about 1 in 500 times, the debris from the long-lived mode consisted of just two pions instead of three. Theory implied that this would be impossible if CP symmetry were valid; the +1 and −1 bookkeeping would forbid it. The implication was that the bookkeeping failed: CP is *not* an exact symmetry of nature. Put another way, it means that in any beam of these kaons, there will eventually be an excess of matter over antimatter at the level of 1 part in 300. This is very tiny but nonetheless real and highly significant. It was the first, and until the end of the century, the only direct sighting of an imbalance between matter and antimatter, other than the (indirect) evidence of our own material existence, 15 billion years after an initially symmetric, balanced, universe of matter and antimatter. In some sense the universe is like Escher's geese rather than horsemen.

Perfect symmetry: back to the future

So, what are the balance sheets of nature? Physicists seeking the symmetry of the universe will not be happy until everything is accounted for and the sums balanced. They once believed that mirror symmetry was inviolate. Having been proved wrong, and then thinking that this asymmetry would be balanced out by the charge mirror—the combined CP operation (as with Escher's horsemen)—they found a discrepancy in the ledger here too. Though even today we do not yet fully understand how this subtle asymmetry arose, it has given the clue to the true symmetry of the universal fabric. It is this.

Take a particle, replace it with its antiparticle and look in a mirror—this far is the CP operation and the books don't quite balance—but now also *reverse the direction of time*. If one were to do this for any interaction involving particles, the recipe should produce a result that is indistinguishable from the original. The total package of C, P, and T (for "time"), as these interlocking pieces are known, should preserve the patterns of the universe. That the symmetry of the Creation involves C, P, and T reversal may sound arcane, but it is nonetheless one of the most profound fundamental truths that we know. If nature should turn out not to respect this symmetry, then everything that we understand, our theories, and the sum of all experience would require us to go back to the drawing-board.

This is bizarre but not as weird as some cultists like to suggest. For example, it does not mean that antimatter worlds are somehow coursing through our present, emerging unseen from the future, complete with antialiens that are growing younger by the antiday. Imagine living in a universe like ours but where C, P, and T had all been reversed. Whereas our universe is expanding from the original Big Bang, the CPT-reversed version would be hurtling towards a "Big Crunch." Some physicists even suggested in 1999 that in such a CPT-reversed universe, any antipeople would remember what we call the future while having yet to experience what we know as the past. They would regard the "backward" flow of time as naturally as we do the forward.

I do not want to enter the labyrinth of what time is; I am more than happy with the whimsical remark that time is what is needed to stop everything happening at once. For bulk matter, including living things, time is an illusion involving the laws of chance as applied to large numbers of atoms. The CPT perfect symmetry applies to individual atomic particles, the basic players in nature's scheme. The Brookhaven team had discovered in 1964 that CP symmetry is spoiled; so to keep things in balance, the symmetry of time must also fail. The accountants suspected that this must be so, but more than 30 years were to pass before the audit was completed. Finally at CERN, in 1998, the T symmetry was indeed found to be broken and in such a way that the total CPT symmetry of the universe survives. The evidence showed up with the kaons, the same entities that in 1964 had revealed the failure of CP symmetry.

By 1998, technology was sufficiently advanced that it was possible to make kaons, or antikaons, in experiments at CERN and then to watch them carefully throughout their brief lives. The magic of quantum mechanics implies that kaons oscillate into antikaons, and vice versa, as they travel. The question was whether kaons and antikaons are transformed one into the other at the same rate. If time were perfectly symmetrical, the two rates should be exactly equal. To understand why this should be so, imagine filming a kaon turning into an antikaon and then playing the film in reverse. You would see an antikaon turning into a kaon. If the rates differ, then you will have discovered an absolute way to distinguish the real event from a time-reversed film. This is, in effect, what physicists at CERN managed to do in 1998. The multinational team, working together over several years, eventually showed that antikaons turn into kaons slightly faster than the reverse process of kaons changing into antikaons. Thus there is an intrinsic directed march of time even at the level of the basic particles. What was also important was that their measurements, combined with the subtle failure of CP symmetry discovered 30 years earlier, counterbalanced nicely such that CPT overall is a good symmetry of nature.

The failure of T symmetry, where kaons turn into the "anti" version faster than the "anti" versions leak back, implies that if you start with an equal mix of kaons and antikaons, eventually a small excess of kaons—of matter—emerges. Now, in the short life of the kaons this imbalance is trifling and, of itself, not enough to explain the huge dominance of matter in the universe today. Nonetheless, it is proof of such an asymmetry.

The results from decades of study with ever increasing precision, suggest that there may be another place in nature where this matter–antimatter asymmetry takes place, and much more dramatically than in the case of the strange kaons. In 1977, the first examples of "bottom" particles were discovered. Previously unknown, it turned out that these are in effect heavier versions of "strange" particles, such as the kaons; it is the latter that revealed the subtle CP and T asymmetries and so the bottom (or "B") particles should as well. But being so much heavier than their strange counterparts, B particles should show a characteristic and large asymmetry between their matter and antimatter versions. As Bs were abundant in the first moments of the universe, it may be that they contain the secret of how the lopsided universe, where matter dominates today, has emerged.

To unearth the secret, the strategy is to make billions of these ephemeral B particles, and their B antiparticle counterparts, and to study them in detail. To do so a veritable "B factory" has been built at Stanford in California. It is a relatively compact machine on the scale of modern particle physics (being only a few hundred metres in circumference) but involves high-intensity beams of current controlled with a precision greater than ever achieved before. The energy of the beams is specially tuned so that the conditions for creating the Bs are optimized. It has taken years to build the special detector, known as "BaBar," that will record the results and for months its components were thoroughly tested. All that remained was for the counter-rotating beams of the accelerator to be switched on, to clash into one another successfully such that Bs are produced in vast numbers, and for BaBar to record the aftermath.

This potentially historic experiment began in 1999, and confirmed that there are large differences in behaviour of B particles and their B antiparticle counterparts. This is intriguing, but we remain ignorant about the sharp details of how the universe managed to avoid destruction. The final answers may only become known when we peer deeper into the moment of Creation with the most powerful instrument ever built: the Large Hadron Collider (LHC) at CERN. This could be the final act in Lucifer's Legacy on which the curtain rose in 2010. It is still early days at the LHC. The hope is that the experiments will reveal the long-lost perfect symmetry of the Creation and in so doing expose the culprit that has hidden it from view, enabling the structures essential for our present universe, both inanimate and living, to have emerged.

Chapter 13

Lucifer's legacy

A colleague of mine, who was formerly a student at Edinburgh University in 1964, returned from his summer vacation that year and found a note from his research supervisor awaiting him. It read "This summer I have discovered something that is totally useless" and was signed "Peter Higgs." Forty years later, at CERN, scientists from over 80 countries prepared to start experiments on a project costing the equivalent of over a billion dollars. One of its central aims is to test Higgs' "useless" idea. In 2012 the discovery of Higgs' boson gave the first evidence that the idea is correct.

Theorists have brought together the regularities among the forces and among the basic particles on which they act, and created grand unified theories whose underlying equations are as beautiful to mathematicians as are the works of Shakespeare to literati. Historically, beauty in the mathematics has often been a guide to reality in the past (witness Dirac's equation that uncannily "knew" of antimatter in advance of its discovery) and the theorists suspect strongly that the same will be the case this time. There is, however, a problem. These fundamental equations have perfect symmetry only if all the particles, both of matter and the force carriers, have no mass at all. The real universe as we know it is not like this: mass makes the world go round.

Isaac Newton in the seventeenth century showed that weight is equivalent to mass, and thereby developed his universal law of gravity. Two hundred and fifty years later Albert Einstein showed that mass is equivalent to energy ($E = mc^2$). Among a host of implications, this revealed the awesome power of the atomic nucleus and the conversion of matter and antimatter into energy (and vice versa) which at LEP (see page XX) has opened our modern view of Creation. Although Newton and Einstein provided these great insights, neither of them knew what mass actually is. The ideas to which Higgs' name has become associated explains the origin of mass and so suggests that mass may be what disturbed the perfect symmetry that was the aftermath of Creation: it was the entry of mass that gave substance and form to the universe.

The ideas of previous chapters, namely that nature can hide symmetries, generating patterns and structures in the cold from more pervasive symmetry in the heat are now applied to the entire universe. The beautiful symmetric equations apply only at extreme temperatures—as in the moment of the Creation. Their solutions in the cold, which is the universe now, do not show the primeval symmetry. This need be no surprise; the thrust of the previous chapters has been to demonstrate that nature can hide symmetries. This realization means that there is the possibility of deep connections among apparently unconnected phenomena, and that it is mass that is hiding the true unity.

Mass has become the prime suspect in the mystery of Lucifer's legacy; mass is the spoiler of the symmetry of the Creation and the source of all structure, pattern, and asymmetries since.

The more one looks, the more one realizes the pervasive influence of mass in the basic workings of the universe. The electron is an essential piece of all atoms, yet nature has made two seemingly redundant copies of it (the muon and tau—page 218) distinguished only by their masses. There are also three varieties of neutrino (see page 218), making a total of six particles in the electron and neutrino family (the "leptons"). There are six varieties of quark too; three with positive electrical charge (the up, charm, and top) and three with negative (the down, strange, and bottom). Is there a profound connection demanding that there are six of one and half a dozen of the other, or is it merely a coincidence?

Fig. 13.1 Peter Higgs and an illustration of a part of his theory on the blackboard.

What does appear to be true is that were it not for their different masses, the three sets of leptons or the three sets of quarks would appear identical. For example, the up quark (which is a part of the proton) is identical to the charmed quark, except that the latter weighs more than a hydrogen atom. The third member in this trio, the top quark, weighs nearly as much as an atom of gold. How nature concentrates such an energy into a volume smaller than we can yet measure is but one of the mysteries associated with mass.

The standard model of particles and forces exhibits other patterns, symmetries, that have been hidden by mass. For example, the strengths of the electromagnetic and weak forces are intrinsically the same but their effects differ because of mass. The quantum bundle of electromagnetic radiation, the photon, has no mass and flies across space at the speed of light. By contrast, the analogous bundles of weak radiation, the W and Z, weigh more than atoms of iron. These great masses impede their creation and limit their spheres of influence, making the weak force truly feeble—at room temperature. However, as we have seen, at the extreme energies accessible at LEP the sluggish effects of mass are neutered and the effects of the W and Z are revealed in their true glory. Had the deleterious effects of mass not enfeebled them at lower temperatures, such as those in the sun where they play essential roles in its fuel cycle, then life on earth would have never been.

In previous chapters we have led up to the perfect balance between matter and antimatter that was destroyed in the "Great Annihilation," enabling a matter-dominated universe to have a chance. However, without mass that would have been the end of it: there would have been no atoms, no structures, no life. That we are here is dependent upon the delicate pattern of masses for the basic particles, as the following examples illustrate. (A detailed and more technical version of how the parameters of the Standard Model affect everyday life can be found in Robert Cahn's article in *Reviews of Modern Physics*, volume 68, pages 951–9, 1996.)

At the centre of each atom is its bulky nucleus consisting of protons and neutrons. The neutron is slightly more massive than a proton, which is the source of the neutron's instability and, conversely, of the stability of hydrogen. This is all a result of the down quark being slightly heavier than the up variety. We don't know why nature selected these values. Had it changed them very slightly, so that the up quark was heavier than the down, then the proton would have been heavier than the neutron, and in consequence unstable. Hydrogen nuclei would have decayed and positrons been emitted. These would encounter electrons and their mutual annihilation would pollute the universe with potent gamma rays. Neutrons, by contrast, would have been the stable end products of Creation. The stars, which for us are made of protons, would instead be made of neutrons and the pattern of elements produced in the stellar "cookers" would have been very different from those that we are used to.

At the periphery of atoms are electrons which appear to be fundamental entities. Where their mass comes from, and why it is the magnitude that it is, are also fundamental questions that relate to our existence. The electron in hydrogen encircles the central proton at a distance governed by the electron's mass: were the electron heavier than it actually is, atoms, and ultimately all matter, would be reduced in size. Furthermore, a heavier electron would stop some of the phenomena of radioactivity and transmutation of the elements, with the result that the emergence of the essential elements would never have happened. Conversely, were electrons lighter, these radioactive transmutations would still occur but the ensuing matter would be utterly changed.

If one could turn a dial that reduced the electron mass gradually to zero, one would be aware of atoms becoming larger as the electron mass falls. However, were the electron mass to be exceedingly small, or ultimately zero, the atoms would become so distended as to have no recognizable independent existence. Cold matter would be plasma, a swirling gas of electrically charged particles, as in the (hot) sun of the actual universe. The universe would be a "soup" of radiation and particles but with no solids, no structures, and no elements suitable for life.

These delicate balancing acts are true not only for the basic particles of matter but also for the quantum bundles of the force fields, such as the photon and W and Z. As we have already noted, the masses of W and Z control the strength and effects of the weak force. The W is at work in a weak interaction process that is essential to our everyday lives: the fusion of protons that fuels the sun. The Z determines some aspects of supernova explosions that pollute the cosmos with the elements needed for planets, plants, and all living things. We are stardust or, if you are less romantic, nuclear waste. Were the mass of the Z different, then the supernova sums would be disturbed; whether this would have ruined our chances is moot but it is certainly true that the mass of its sibling, the W, is critical for us.

Had the W been twice as heavy as it is, the weak force would have been even weaker than in reality. In such a circumstance the nuclear processes at the sun's heart would be enfeebled and its hot centre would be less able to resist the weight of the outer layers pressing upon it. The sun would shrink and its surface temperature rise from its present 5800° to 7000°. The beauty of the rainbow would still occur but beyond its blue horizon, in the ultraviolet, the intensity of the radiation would be much greater and the evolution of green plants would be drastically disturbed. The ability to resist skin cancer would have been one of the tests of evolution in such a world. Though we might have emerged in such circumstances, the other choice, of a lighter W, would almost certainly have spelled our doom. In this case it is not that we would have died prematurely: we would never have been born.

As I said earlier: we are stardust. Supernovae have cooked their original hydrogen into the heavier elements—in particular the carbon, nitrogen, and oxygen—that are the templates of amino acids and life. Collect enough atoms of these elements together and they become self-aware. How this happens we don't understand, but we do know that it takes a considerable amount of time: over four billion years of sunlight shining on the earth was needed before intelligence at the human scale emerged. Had the W been lighter, the sun would have burned faster. It is currently halfway through its life, and we are here; had it burned twice as fast, it would have expired before the most primitive life forms had begun.

Had the W and Z been massless, like the photon, then the universe would have been a peculiar place. Electrically charged particles, such as electrons, would probably have been permanently entrapped by the electroweak forces in the way that quarks appear to be permanently confined within protons and neutrons in the real world. As the gluonic radiation of the quarks' force fields is stuck within the protons and neutrons, so might electromagnetic radiation have been imprisoned within atoms. The physics of such a universe is for science fiction writers to explore.

These alternative worlds, where the neutron is lighter than the proton or where the electron mass is marginally larger or smaller than in reality or where the carriers of the forces were lighter or even massless, show how things might have been. Although we understand a great deal about the physical laws that determine how the particles and forces cooperate to create the world that we inhabit, we do not know why the world is this one and not one of the many that "might have been."

There is much dreaming of a final, grand, unified theory, but at present the values of the masses of the elementary particles have to be inserted into our calculations at appropriate moments without any understanding of why they have the "magic" values that they do, or even why they have any masses at all. By considering what would be different if some or all of these masses were changed even slightly from their real magnitudes, we see how delicately balanced existence is and we realize how much we really have to explain. These questions have been with us ever since the muon (a heavy version of the electron) was discovered in 1947 and caused the Nobel physicist Isidor Rabi to exclaim "who ordered that?". What is special today is that this and similar questions (such as the reasons for the masses and existence of the "superfluous" copies of the electron and quarks) are graduating from the realms of philosophy into those amenable to scientific examination. The goal of thousands of physicists, currently gathering at CERN, is to answer these most fundamental questions. To set the scene for what they plan to do, I will first give an outline of what Peter Higgs' "useless idea" implies.

The "Higgs' field"

To test Higgs' idea would require the largest and most expensive scientific endeavour ever. In 1993, as governments debated this, the British Science Minister, William Waldegrave, set the physicists a challenge. In order that he present their case in the best way during discussions with other cabinet ministers, he suggested that the physicists should help him by giving him a concise explanation of the Higgs' idea. In order to stimulate them, Waldegrave offered a bottle of vintage champagne for the best attempt. However, and here was the real challenge, as the attention span of busy politicians dies rapidly after a page, the description must fit on a single side of A4 paper.

The idea that one can do such a thing is dangerously appealing. Interviewers in the media, most often with little or no scientific awareness (which is still regarded with apparent pride by one well-known British radio personality), seek instant explanation. Failure to provide such is then taken as proof, if any were needed, that scientists are out of touch with reality; a scientist unaware of the arts is a philistine but the converse, apparently, can be acceptable. The expectation that a lifetime of study and sweat can be pithily condensed for a media personality is to belittle the dedication that made the expert worthy of being asked in the first place. Mr. Waldegrave's challenge was not of this type but I was concerned nonetheless.

I was reminded of Richard Feynman who, upon winning the Nobel Prize for physics, was asked by a local television reporter to describe the work for the viewers "in less than 20 seconds." Feynman retorted that if it were possible to have done so, then the work

Fig. 13.2 Cartoon sequence inspired by David Miller's explanation of Higgs' theory. In (a) the party workers represent the Higgs' field; in (b) a famous politician enters; and she attracts clusters of workers around her (c).

Fig. 13.2 (*continued*)

would not have been worth the Nobel Prize! I recalled this story in the letter that accompanied my own response to Waldegrave's challenge. Rather pompously I added that I was concerned at being asked to describe such profound ideas on a single sheet of paper, but nonetheless I would be prepared to face the challenge if he, in a similar amount of space, would describe the Maastricht Treaty. Neither he, nor I succeeded but Professor David Miller of University College London won a bottle of champagne for his entry.

If you want to capture your readers' attention, identify with them. David Miller's winning entry cleverly used a political analogy. First, he imagined a cocktail party attended by members of Waldegrave's own Conservative Party. They are uniformly spread across the floor, each talking to their nearest neighbour. At this point, the former Prime Minister, Margaret Thatcher, enters and proceeds to walk across the room. All of the people in her immediate vicinity are strongly attracted to her and gather around her momentarily until, as she passes, they detach and return to their even spacing. The knot of acolytes impedes her, slowing her progress. Once moving she is harder to stop, and once stopped she is harder to get started again as the clustering has to be reconfigured. Her inertia is greater: in collusion with this knot of acolytes, she has a greater mass than normal.

In three dimensions, and with the complications of relativity, this is the essence of Higgs' idea. The masses of the particles come as a result of them interacting with a field that permeates space— the party workers in Miller's analogy. This field, the "Higgs' field," becomes distorted locally when a particle passes through. It is this distortion, as the field clusters around the particles, that gives the particle a mass. To further Miller's analogy: had the leader of an opposition party entered the room instead of Mrs. Thatcher, he might have been ignored and passed through unimpeded, like a photon.

While Miller's analogy neatly illustrates the essence of Higgs' ideas, it does not touch on issues such as how temperature or energy plays a role, nor why the LHC is needed to prove the thesis.

To begin to answer these questions, first recall two basic facts that we have met in recent chapters: physical systems when left to their own devices seek to attain a state of lowest energy; to do so they may change their phase, as when liquids freeze, giving up energy as heat, or when the individual magnetic atoms in a lump of iron line up and imprint magnetism on the metal. Armed with these, to introduce Higgs' idea in our universe I will first recall some of the things we have met concerning the nature of magnets.

In iron and other metals, each electron is spinning and acts like a small magnet, the direction of its north–south magnetic axis being the same as its axis of spin. For a single, isolated electron this axis could point in any direction: this is an example of the rotational symmetry of three-dimensional space where any direction is as good as any other. However, when they are in iron, neighbouring electrons prefer to spin in the same direction as one another, as this minimizes their energy. To minimize the energy of the entire crowd, all of them must spin in the same direction; each little magnet points the same way and it is this that becomes the north–south axis of the whole magnet. Here is a classic example of hidden symmetry in that the laws of physics for spinning electrons have no preferred direction, whereas the ensuing magnet clearly does—the direction connecting its magnetic poles.

That is how things are in the cool, but heat the iron above 900°C and the extra energy that the heat provides will be more than enough to liberate each of the spinning electrons from the entrapment of its neighbours. The spin axes of these mini-magnets will point in random directions once more; the rotational symmetry is restored and the magnetism disappears. Cool the metal again and magnetism will return, but the direction of the new north–south axis will most probably be different to what it was before. The fundamental symmetry of space treated all directions equally; the solution at low energy hides this symmetry by having chosen a special direction—the north–south axis of the magnet—at random.

Instead of heating the magnet to liberate the spins entirely, suppose instead that we stay in the cool, magnetic phase and give the spins a small pulse of energy so that they wobble. It is possible to make the direction of the spin—the local magnetic north in effect—vary in a regular, wave-like way, from point to point. These are known as spin waves. Just as electromagnetic waves are bundled into quanta called photons, so spin waves are bundled into "magnons." These magnons act like particles, much as photons do.

Imagine for a moment what our experiences might have been if we were mythical creatures that had evolved inside a magnet; we would interpret its unchanging features as "empty space." In developing our first descriptions of our "universe" we would be unaware of rotational symmetry because we would experience a preferred direction to space, namely that due to the ubiquitous magnetism throughout. Asymmetry is what we would regard as the natural order.

When it comes to asymmetry in the land of the blind the one-eyed is king. After hundreds of years, one scientist among the mythical creatures realizes that "empty space" is not really empty but that it has a structure—what we with our greater wisdom recognize as the magnetic spinning electrons. Furthermore, the scientist realizes that it is this structure, combined with the fact that the creatures are living below 900°C, that has given rise to asymmetry: the underlying laws are truly symmetric. At first this is just an idea but the creatures discover that they can do an experiment to test it: heat the "vacuum" to above the critical temperature and the true symmetry will be revealed. Alternatively, energize the "vacuum" enough and magnons (the evidence of structure in "empty" space) will be seen. All that is lacking is the technology needed to energize a small portion of space to the critical level.

It turns out that these creatures might not be so mythical; they could be us, as we suspect that this is an allegory for our real universe. Higgs' theory is built on a perception that the "vacuum" is really a structured medium and the magnons in our universe become particle manifestations of the structured vacuum; they are known as "Higgs' bosons." Discovering Higgs' boson is the first direct evidence that there is an all-pervading field giving fundamental particles and forces their characters.

So, having had Miller's analogy and the magnet example to prepare us, let's imagine ourselves not as creatures whose "vacuum" contains unseen magnets but as our true selves in a universe whose vacuum contains an unseen Higgs' field. Recall for a final time the two basic rules: physical systems, when left to their own devices, seek to attain a state of lowest energy and to do so they may change their phase. Armed with these two mantras, imagine pumping a container empty and lowering its temperature as far as possible. You would reasonably expect that by doing so the volume within the container would have reached the lowest energy state possible since upon adding anything to it, you would necessarily be adding the energy associated with that stuff. However, the "Higgs' field" has the bizarre effect that when *added* to the container it *lowers* the energy still further, so long as the temperature is less than the critical value of $10^{17}°$.

The universe is far colder than that everywhere today and has been since a mere 10^{-14} seconds after the Creation. Ever since that first moment, when the universe cooled below $10^{17}°$, the Higgs' field has been locked into the fabric of the universe.

This idea sounds quite unreal: it implies that all of space, in and around us, is filled with this Higgs' field. So why are we not aware of it? The answer is: we are! All structure, existence, and being are testimony to the Higgs' field having frozen mass onto the fundamental particles. This is what theorists believe has happened.

Phase transitions and changes in symmetry go together, as we have seen in Chapter 9, and the Higgs' example is no exception. In this case it is the symmetry of the particles of matter and the carriers of the forces where the change is felt. In the hot phase there is a supersymmetry, where all of the particles of matter and the force carriers are united. As the freezing of water selects six directions for the ice crystals to develop out of the total rotational symmetry that had existed previously, so patterns emerge among the particles as the universe cools below the critical Higgs' temperature. One aspect of these patterns is that particles gain masses and thereby their distinct identities.

These ideas are compelling but, as we must keep reminding ourselves, they are only pieces of mathematics written on paper and the challenge is to test them experimentally to see if the universe really did begin this way. The mythical creatures in our magnet example did so by creating localized hot spots at 900°; for us the challenge is similar, except that the required temperatures approach 10^{17}°! As the electromagnetic field is manifested in quantum bundles, the photons, and as spin waves in magnets form magnons, so should the Higgs' field have particle bundles, the "Higgs' bosons." According to the theory, and in a way that almost begs the question of which came first, the chicken or the egg, these Higgs' bosons are themselves given mass by interaction with their own field.

A critical question for the planned search was where to look—specifically, how massive is the Higgs' boson? Its mass was narrowed down within a range of possibilities and experiments were planned to cover this range in order to find it. We knew that it must weigh over 100 GeV (that is approximately 100 times the mass of a proton) as otherwise it would have been seen already at LEP. At the other extreme, if it were heavier than about 600 GeV it would affect the behaviour of known particles in ways that experiments would have already detected. In some more recent developments of Higgs' theory, more than one boson is to be found. There was excitement when theorists realised that one of them could have been lighter than 200 proton masses and be discovered during the final year of LEP's operation or at Fermilab in Chicago. It didn't turn out that way, however. To explore the complete range up to 600 GeV required the Large Hadron Collider (LHC) at CERN ("hadron" being a family name for the protons and atomic nuclei that will be used in the collider).

The LHC

Whereas LEP used lightweight electrons and positrons, the LHC swings two counter-rotating beams of protons around the same 27-kilometre circuit that was previously LEP. The protons are some 2,000 times more bulky than electrons and pack a greater punch—capable of reaching to higher energies, earlier epochs, than can LEP.

However, the magnets that guide the electrons in LEP are too feeble to steer the bulky protons. To keep them on track needs stronger magnetic forces than had ever been used in a CERN accelerator. So LEP's magnets were removed and an entire new set installed in their place. Powering these magnets posed new technological challenges for which superconductivity is the key. Superconductivity is the ability of some materials to conduct electricity without resistance or energy loss; this reduces the power demands of the LHC. However, superconductivity makes demands of its own: for the magnets to operate as superconductors requires temperatures so low that the LHC has to operate about 300° below room temperature—colder even than outer space. A superconducting installation at such temperatures throughout a length of 27 kilometres had never been built before. Each magnet is over 14 metres in length and more than 1200 of them were needed to complete the circuit.

In LEP, the beams of electrons and positrons had opposite electrical charges and so a single ring of magnets was sufficient to guide them around the circuit in opposite directions. The LHC by contrast accelerates two beams of identical protons in opposite directions and to do so requires two rings of magnets. So LHC is effectively two accelerators in one. To keep it compact and as economical as possible, specially designed magnets with two pipes, one for each beam, have been built. These bring the protons into head-on collisions, similar to those at LEP, but with far greater violence and with much more debris to be recorded and analysed.

The detectors at LEP would be totally inadequate and were super-ceded by behemoths that dwarf even the LEP giants.

One of the detectors is a 12,500-tonne monster known as CMS, the MS referring to its technical nature as a magnetic spectrometer while the C stands for "compact"! While it might stretch credibility to think of such a giant as compact, indeed it is relative to the other detector, the aptly named ATLAS, against which it pales (the name, however, is more mundane, ATLAS standing for *a t*oroidal *L*HC *a*pparatus, in reference to its doughnut shape). If the magnitude of CMS is compared with a cruiser, then ATLAS is like a battleship. Twenty metres tall, it is double the height of the LEP detectors, and akin to the size of an office block. The dimensions of the detectors at the LHC are such to enable them to track the debris—ejected from collisions at energies far more extreme than those that were the norm at LEP. To control this debris, which is moving faster than anything ever seen before, and then to tease out its secrets, requires magnetic fields in the detector that are at the limits of technology. The energy contained in the magnetic fields of CMS, for example, would be enough to melt 18 tonnes of gold.

The protons in the LHC are accelerated in bunches that pass through one another 40 million times each second. Every time two bunches cross, some 20 individual collisions between pairs of protons will occur, making a total of some 800 million collisions each second. With only a billionth of a second elapsing between successive collisions, debris from one is still travelling through the detector while the next collision is already taking place. To record what happens requires information to be processed at a rate that is equivalent to everyone on earth making 20 phone calls simultaneously. In ATLAS, the innermost sensors alone contain more than 10 thousand million transistors—a similar number to the multitude of stars in the Milky Way. The technological challenges have pushed the frontiers of research and development in diverse fields such as information processing, in the production of detectors that can survive and operate successfully for long periods while subject to intense radiation, and in superconductivity. For many engineers, the project itself has defined the cutting edge of technology; the discovery of the Higgs' boson is a bonus. For physicists this is not the beginning of the end but the end of the beginning. Having found evidence of the Higgs' boson the challenge now is to understand exactly how it behaves, and why the particles and forces have their specific properties.

Supersymmetry and the fifth dimension

We began our journey into symmetry 11 chapters ago with gravity and, you may have noticed, have said hardly anything about it since. Gravity is not in the standard model of particles and forces. It is so feeble between individual atoms that it is irrelevant for experiments in particle physics, but it has to be included in a final theory. There are some tantalizing hints on how this might be achieved, and that may lead to profound surprises at the LHC. These touch, in part, on the belief that there may be one more aspect of symmetry at work among the particles and forces than we have mentioned so far, and that this could provide the key to completing the picture of symmetry and its disappearance. Furthermore, it may begin to teach us how gravity, mass, and the other forces are united. It has even led physicists in the final year of the twentieth century to contemplate seriously the idea that there may be a fifth dimension, and that it could be revealed at the LHC.

This final link is known as "supersymmetry" (colloquially known as SUSY) and is a marriage of two symmetries that we have already met. On the one hand we have identified a clear symmetry among the leptons and the quarks; on the other we have discerned a clear commonality among the carriers of the forces. As the forces are transmitted by particles and, in turn, the particles feel forces, the question that suggests itself is whether the particles of matter—the leptons and quarks—are related to the force carriers. The premise of SUSY is that they are.

The idea that all of these entities were united in the original heat of the Big Bang and became differentiated by their masses as the universe cooled is remarkable and adds to the list of delicate chances that enable us to be. The matter particles and the force carrying ones that we have so far discovered are different in an essential way—in the rate that they spin. Electrons spin at half the rate of a photon and so are defined as having spin 1/2; all leptons and also the quarks spin this way: collectively they are called "fermions" (after the famous Italian physicist Enrico Fermi). Contrast this to photons—their spin in these units is one; the same is true for W and Z and also gluons. Anything with an integer number for its spin is collectively called a "boson" (named after the Indian physicist Satyendra Bose). Hence the Higgs' boson, which is predicted by Higgs' theory to have no spin at all, thus to have an "integer" spin of zero.

SUSY connects these fermions and bosons—on paper. If these pieces of mathematical poetry are realized in nature then there exist whole families of particles of fermions and bosons awaiting discovery. Originally, at the start of time, they and the particles we know were united; as the universe cooled, the onset of mass hid this twinning such that the quarks and leptons ended up lightweight, whereas their supersibling bosons (known as squarks and sleptons) were left with masses (energies) similar to those characteristic of the energy at which the Higgs' phenomenon is manifested.

If the universe really is like this, then we are indeed fortunate that the onset of mass made the leptons and quarks lightweight, rather than the squarks and sleptons. If the concrete world had been made of bosons rather than fermions, sleptons rather than leptons, then no atoms could ever have emerged, let alone anything like us. While life, and indeed all biological and chemical activity, depends upon the ability of molecules to combine and transform, the stability of matter and the richness of chemistry is because some things do NOT combine. Profound aspects of the quantum theory show that this is because the basic constituents of atoms, of all matter, are fermions—have spin 1/2. Fermions are like cuckoos (where two in the same nest is one too many) whereas bosons are like penguins (the more the merrier). Laser beams are an example of this, where large numbers of photons (bosons) act in concert. Contrast this to what happens when you stand on the floor—the atoms in your feet are mostly empty space as are those in the floor, yet you do not sink through. The resistance of atoms to passing through one another and to coalescing into structure-less forms is intimately due to the "cuckoo" nature of their basic fermion constituents: an electron in the seat of your chair will not welcome the intrusion of an electron that is in the seat of your pants—solidity ensues.

However, had electrically charged bosons been the lightest par-ticles in the universe, the resulting molecules would have merged into a featureless blob. In turn, the mutual attraction of one blob for another would be so great that they too would merge and end up smaller than the originals. Such a world, where the squarks and sleptons were light, would have its matter collapse like something from a ghastly horror movie. So if SUSY really was part of the original symmetry of the Creation, we must be grateful that the breaking of the symmetry gave light fermions rather than light, electrically charged bosons. Whether indeed SUSY was the origi-nal unity in Creation and, if so, how it was that the fortunate out-come was chosen as the universe froze, is another question for experiments at the LHC.

In the course of probing the mathematics of supersymmetry, theorists appear to have stumbled upon the possible long-sought "theory of everything," known as "superstring theory." The super-string theory comes close to being impossible. It has proven so difficult to formulate a theory of gravity that is consistent with quantum theory and relativity, that the discovery of the super-string theory, which does so, has encouraged theorists to suspect that this is indeed the unique true description.

Among its many bizarre implications is that all of the natural forces are actually gravity in higher dimensions. At the dawn of the twentieth century the three dimensions of space were the rule. Then Einstein married space and time and gave us the modern four-dimensional picture where gravity is the result of curvature in the fabric of "space–time." If the superstring theory is correct, then in addition to up–down, front–back, and sideways, there could be further directions "within." These higher dimensions are believed to have shrivelled up at the Big Bang while the others have grown with the expanding universe. Until very recently it had been thought that the scale of these dimensions is so small that we are likely to remain unaware of them until well into the twenty-first century, if ever. But there are now some tantalizing suggestions that one of these, the "fifth dimension," could be almost within sight. It is possible that space–time is like Ementhal cheese, with little bubbles permeating it, whose sizes are at the edge of our present abilities to measure.

If SUSY is indeed proved to be part of nature's scheme then the superstring theory, where all the forces are gravity, gets its odds shortened radically. Historically, gravity has been thought to become strong, and ultimately unified with everything, at energies of 10^{19} GeV—the so-called "Planck energy." Higgs' boson is revealed at below 10^3 GeV, a value that has no immediate connection with the Planck energy. However, some theorists examining the fine details of what has been learned, and of the mathematics of the underlying theories, discovered a loophole. If there is a fifth dimension that is curled up on length scales that are on the borderline of what we can measure, then the Planck scale turns out to be much less than we had previously imagined in our four-dimensional perspective. It is even possible that the source of gravity, of the fabric of space–time, becomes apparent at essentially the same energies, or temperatures, as the Higgs' phenomenon.

This possibility is now exciting many theoretical physicists. The tantalizing prospect is dawning that maybe at the LHC we will be able to see for the first time the ultimate foundations of reality. The long-sought unification of Einstein's theory of gravity with quantum theory, the understanding of all the forces as the "smoking guns" of higher dimensions, and the realization of the original symmetry of nature might be just around the corner. The idea of higher dimensions has fascinated philosophers and science fiction writers for decades; now they are being regarded by many physicists as things whose consequences we will be able to measure.

Like the mythical creatures living inside a magnet who suddenly had the existence of the magnet revealed to them, and with it enlightenment as to the true reality, so can we anticipate some joyful revelations from the LHC. And when the true symmetries of the Creation are finally known, the way that nature broke those symmetries will also become clear. For it is this that is the origin of our existence. The Creation of a perfect symmetrical universe, had it stayed that way, would have prevented any intelligent life ever being able to know it.

A collection of atoms of carbon, nitrogen, and a few other elements can be self-aware, can look out into the heavens, can build machines that ask questions of reality—none of this could have happened but for asymmetry. The asymmetries that emerged when the universe froze are what have enabled life to be. Why they are as they are is the greatest mystery of why we exist. There may indeed be a singular source of all asymmetry, the reason for form and existence. My fascination began in the Tuileries Gardens with the headless statue; this book, if not its theme, is therefore Lucifer's legacy.

Epilogue

Three years later, on another beautiful day in another spring, I returned to the Tuileries. It was as if nothing had changed. The trains still meander through southern England; northern France remains a place to pass through at high speed; Paris in the spring is timeless. The Tuileries hosted lovers as always, walking along its symmetrical paths, admiring the ordered flower-beds, pausing at the statues. But something was different: the headless Lucifer was no more—two identical, complete, diabolic twins now faced one another across the centre line of the park. The devilish asymmetry had been repaired. Had I come to see the gardens today, rather than three years earlier, it would not have offered me the essential hint that had set me on my quest.

But the clues are there if only you look. And the question had already been asked, half an hour's walk away today but 150 years earlier in time.

Leaving the gardens on the south side, I crossed the River Seine and entered the alley-ways of the left bank. This brought me past the Grand Ecoles and the Academie des Sciences (where 100 years before Henri Becquerel announced his discovery of radioactivity, and with it helped start modern science) and also the College de France (where in 1835 Jean-Baptiste Biot first found that sugars rotate the plane of polarized light). It was Biot's discovery that was the first hint of an asymmetry in molecules that are important to life and Becquerel's that, nearly 100 years later, would eventually reveal an intrinsic asymmetry in nature's fabric. Finally, my journey brought me to the Institut Pasteur. It was Louis Pasteur who first realized that the mirror asymmetries in molecules control our lives.

The first hints of Lucifer's legacy were not in the Tuileries in the final years of the millennium but had been anticipated across the river in 1860 where Pasteur, who knew nothing of radioactive transmutations, of electroweak symmetry, or of Z particles, had made the following potentially prophetic remark: "Life as manifested to us is a function of the asymmetry of the universe and of the consequences of this fact. The universe is asymmetrical. Life is dominated by asymmetrical actions." And then his great insight and challenge, whose truth may soon be resolved: "I can even imagine that all living species are primordially in their structure and in their external forms, a function of cosmic asymmetry."

Index

About the Author

Frank Close is a Fellow of Exeter College, Oxford University. He was formerly Head of Theoretical Physics Division at the Rutherford Appleton Laboratory and was Head of Communications and Public Education at CERN, the European Laboratory for Particle Physics, from 1997–2000. He has been a Fellow of the Institute of Physics since 1991, and was awarded the Kelvin Medal in 1996. He also presented the Royal Institution Christmas Lectures in 1993. His previous books include *The Infinity Puzzle* (2012), *Neutrinos* (2010), *Antimatter* (2009), *Nothing* (2009), *Too Hot to Handle* (1991), *End* (1988), *The Cosmic Onion* (1983), and *Introduction to Quarks and Partons* (1979). He has also written an app for particle physics: The Particles. See: https://itunes.apple.com/app/the-particles/id601382793?mt=8&ign-mpt=uo%3D20